The Great Basin Seafloor

The Great Basin Seafloor

Exploring the Ancient Oceans of the Desert West

Frank L. DeCourten

The University of Utah Press

Salt Lake City

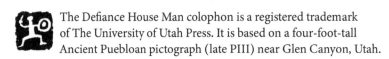
The Defiance House Man colophon is a registered trademark
of The University of Utah Press. It is based on a four-foot-tall
Ancient Puebloan pictograph (late PIII) near Glen Canyon, Utah.

ISBN: 978-1-64769-049-6 (paperback)
ISBN: 978-1-64769-050-2 (ebook)
CIP data for this volume is available from the Library of Congress.

Errata and further information on this and other titles available online at UofUpress.com
Printed and bound in the United States of America.

Contents

Figures

Acknowledgments

Any book that explores aspects of the ancient Earth is a collaborative venture. Our current knowledge of the deep history of land and life on our planet did not materialize spontaneously, or quickly. Instead, it reflects the painstaking field studies and meticulous laboratory research conducted by legions of geologists over the past several centuries. Even in the Great Basin, where formal scientific explorations did not begin until the mid-1800s, we are indebted to the hundreds of geologists and paleontologists who have helped illuminate the fascinating story of the Great Basin seafloor through their studies of the rocks, fossils, and geological structures of the region. I have been fortunate to know, and learn from, some of the most diligent and perceptive of these scientists. Their knowledge, interest, encouragement, and generosity in sharing perspectives were critical to completion of this book, and I am deeply grateful.

As I hope is obvious to readers, this book is an expression of my love for, and fascination with, the bold and sweeping desert landscapes of the Great Basin. This affection began in the 1970s, when my interest in the region was ignited, and later guided, by several scholars who made lasting impressions on a humble geology student and young professional. In particular, Dr. Michael A. Murphy at the University of California, Riverside, Dr. William Lee Stokes at the University of Utah, and Dr. Lehi F. Hintze at Brigham Young University were all inspirational scientists who have helped shape our current understanding of the geological history of the Great Basin region. They were also generous and kind mentors who always took more interest in my education and work than I felt was deserved.

I am also grateful for the assistance of several other individuals and institutions in compiling the information summarized in this book. Dr. Paula Noble, Professor of Geological Sciences, and Garret Barmore, Curator of the W. M. Keck Museum at the University of Nevada, Reno, generously shared information and facilitated access to museum specimens in support of this project. Sali Underwood, Curator of Natural History, Nevada State Museum, Las Vegas, and Carrie Levitt-Bussian, Collections Manager at the Natural History Museum of Utah were cordial hosts on my several visits to examine fossil specimens in their collections. I also thank Bruce S. Lieberman, Senior Curator, Division of Invertebrate Paleontology, Biodiversity Institute, University of Kansas, for allowing use of figured KUMIP specimens. Leslie L. Skibinski, Collections Manager at the Paleontological Research Institution in Ithaca, New York, kindly allowed permission for the use of images from the web portal maintained by that institution. The Utah Geological Survey was very helpful in allowing the use of graphics and maps from several of its publications. Utah artist Carel P. Brest Van Kempen graciously provided his exquisite reconstructions of the Cambrian and Ordovician seafloors as illustrations. University of Utah geologists Paul Jewell and William Perry provided valuable feedback that significantly improved an early version of the manuscript. The meticulous copyediting by Jeff Grathwohl greatly improved the original manuscript. Finally, the publication of this book would have been impossible without the encouragement and help of Reba Rauch, Acquisitions Editor at the University of Utah Press.

This book is accompanied by a brief, supplemental field guide that is available at

https://uofupress.lib.utah.edu/?s=supplemental+field+guide

The guide identifies many accessible localities where the tangible rock record of events described in this book can be explored firsthand. I encourage you to deepen your understanding of the Great Basin Seafloor by visiting these sites and directly examining the evidence of its immense history.

1

Vanished Oceans of the Great Basin Desert

Under the incandescent blue of a western Utah summer sky, I watched the dust devils twist across the searing surface of the Tule Valley hardpan like a tribe of banshees driven insane by the heat. In the distance, the naked ledges of rock in the Barn Hills shimmered through currents of warm air rising above the hardpan's blinding glare. Sitting in the shade of my truck, absorbing this ethereal scene over a cold beer, I suddenly appreciated why the stark landscapes of the Great Basin grip the human spirit so powerfully. Deserts, wherever they exist, are places of visions. And these visions may extend beyond the mere mirages that result from sky-light refracted by heated air currents.

Throughout human history, desert lands have consistently produced more than a few prophets, messiahs, shamans, philosophers, mystics, lunatics, and, of course, writers. It seems that the vast and vacant expanses and uncloaked austerity of arid places prompt us to think deeply and differently about the mysteries of nature, humanity, and our place in the world. No one travels the dusty back roads of the desert, or sleeps under its spangled night skies, without pondering such things. Maybe this tendency to drift into the transcendent, along with the shocking desolation of the surrounding landscapes, is source of the many visions evoked by deserts. After all, is it a mere coincidence that several of the world's major religions originated in desert regions? Reflect also on the prominence of natural imagery in the spiritual-

ity practiced by Native peoples in desert lands and the deep connections between humanity and nature embodied in their precepts. From religious inspiration to the rantings of eccentric renegades to alien abductions, deserts are without question places where the extraordinary becomes normal, whether it pertains to spiritual wisdom, philosophical insight, or even the land itself.

This book is about one of the most intriguing desert visions imaginable: the floor of a vast ocean that submerged the Great Basin region long before the modern dry deserts. The dissonance between the scorched present and the ghost of a sunken prehistoric world is an enchanting paradox. However, this is no metaphysical hallucination, the product of desert meditation, mysticism, peyote, or cold beer (although the latter certainly doesn't hinder the process). Instead, our current knowledge of the seas that submerged the Great Basin for millions of years results from more than 150 years of geological investigation into the solid bedrock exposed in the mighty escarpments and rugged ramparts of the desert mountains of western Utah, Nevada, and eastern California. Since the mid-1800s, when geologists first began swarming this region, the vision of an ancient seafloor has slowly emerged, becoming more vivid, more complete, and more intriguing as each new rock exposure was studied. Applying the fundamental principles of geology to the rocks of the Great Basin, we now have indisputable evidence that

FIGURE 1.1. The Tule Valley hardpan of western Utah, with layered sedimentary rocks of the Great Basin seafloor in the distance.

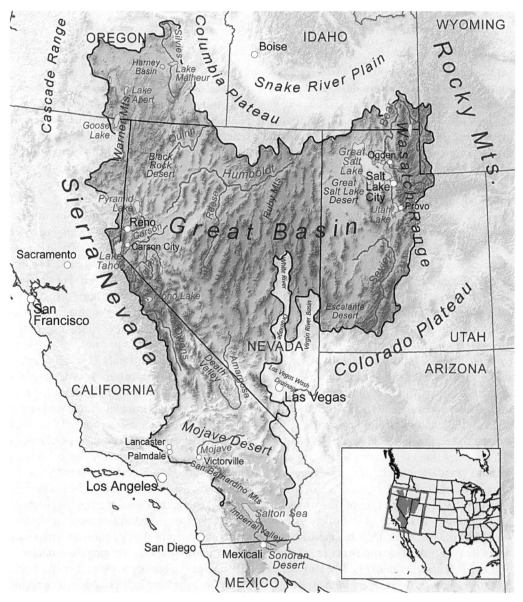

FIGURE 1.2. The Great Basin of western North America. Heavy line denotes the limits of the Great Basin based on internal drainage. Base map by Karl Musser.

the region, however lofty and dry it may appear today, lay almost continuously beneath the waters of vanished oceans for more than 400 million years. The bed of this ancient ocean, where enormous amounts of mud, muck, and sand accumulated during an immense period of inundation, is our subject—the *Great Basin Seafloor.*

Exploring the captivating history of oceanic events recorded in the hardened sediments of that ancient seafloor is a thrill not reserved for geologists only. It is an adventure upon which anyone with a curiosity about the natural world and a pair of hiking boots can embark. All that is necessary to explore the rocks that record the bygone seas of the Great Basin is the interest to do so along with a few basic geological concepts that allow an accurate reading of the rocks. With these few interpretive tools, a person has a lifetime of exhilarating discovery in store among the rock successions exposed on the slopes and in the canyons of the many Great Basin mountain ranges. Among other things, this book is designed to provide those tools and to serve as a guide in your personal exploration of the deep history hidden in the rocks of America's most isolated and vacant natural region.

The intimacy with nature that arises from grasping the profound antiquity beneath the scenery of natural landscapes is one of the greatest gifts of modern science to humanity. Aesthetically, most people can appreciate the stunning vistas of the Great Basin deserts, but geology offers us the ability to see *into* the mountains and to know them deeply. The quest to know mountains, rather than simply viewing them, is a thrilling journey. Let's get started!

Definition and Geologic Overview of the Great Basin

The term *Great Basin* was first used for the mountainous terrain between the Rocky Mountains and the Sierra Nevada of California by John C. Frémont in the early 1840s. Between 1842 and 1854, Frémont made five major expeditions across western North America, several of which crossed the Great Basin in multiple directions. His explorations were mostly mapping expeditions, and, though he did recognize some geological features in the regions he visited, it was primarily the drainage patterns of rivers that inspired his term, still in use for the desert lands of western Utah, Nevada, and eastern California. Frémont was the first to fully recognize that none of the rivers in a vast portion of western North America reached the Pacific, and thus, the region was a "great basin," disconnected from the through-flowing river systems such as the Colorado to the southeast and the Columbia to the north. The drainage divide surrounding the Great Basin—as Frémont defined it—was eventually mapped and served as the boundary of the physiographic and hydrologic province (Figure 1.2).

Following Frémont's initial mapping of the Great Basin, several of the nineteenth century's most illustrious geologists made their way into the region as part of subsequent expeditions to the largely unknown wilderness. During the mid-1800s, the rocks of the Great Basin mountains were first closely scrutinized. Henry Englemann, a geologist in the expedition of Captain James H. Simpson in 1859, performed the first real geological examination of the region. Among other regions of the West, the Simpson expedition crossed the Great Basin twice from Camp Floyd, Utah, to Genoa in the Carson Valley of western Nevada and back. Both the outbound and return routes of the party were close to what would later become the legendary Pony Express trail and modern U.S. Highway 50.

Due to the outbreak of the Civil War, the report of the Simpson exhibition was not published until 1876 (see Chapter 1 references). But the lengthy document included a geological report in which Englemann asserted that, in the mountains of Utah Territory (which then included the modern state of Nevada), "stratified rocks of the Paleozoic age were found extensively developed many hundreds of feet in thickness." The fossils Englemann collected from these rocks, identified by the eminent Smithsonian paleontologist Fielding B. Meek, included the remains of several marine invertebrates. These creatures inhabited the Great Basin seafloor from which Englemann's stratified

FIGURE 1.3. Limestone layers in the Humboldt Range, Nevada, photographed in 1867 by T. H. O'Sullivan during the expedition of Clarence King. Such rocks provided evidence that the Great Basin Desert was once a submerged seafloor. U.S. Library of Congress photo.

rocks originated hundreds of millions of years ago. Englemann's report was the first published recognition that ancient oceans must have once covered the Great Basin region.

In the later 1800s, several other more extensive government surveys of the West added detailed information on the rocks and mineral resources of the Great Basin region. The Geographical and Geological Surveys West of the 100th Meridian, by George M. Wheeler (1869–1879), and the Geological Survey of the 40th Parallel, by Clarence King (1867–1878), were of particular importance in the early exploration of Great Basin geology. Though both great explorations also included regions outside the Basin, they brought unprecedented geological expertise into the sagebrush expanses of Nevada, Utah, and California. Clarence King possessed geological training from Yale, and his survey produced several extensive volumes, published between 1870 and 1878, describing the geology and mineral resources of areas within the Great Basin. Famed geologist Grove Karl Gilbert was part of the Wheeler Survey and ultimately helped produce numerous reports on the geography, geology, paleontology, and botany of the region. The Wheeler Survey enriched the collection of the Smithsonian Institution by more than 43,000 specimens, many of them rock samples and fossils from the Great Basin. By the time Clarence King became the first director of the U.S. Geological Survey in 1879, the broad extent and nature of the sedimentary record of the Great Basin seafloor was well established.

Even as the great western surveys were in progress, the discovery of rich metal ores began to stimulate even greater interest in the rocks of the region. In places like Virginia City,

Eureka, and Pioche, Nevada, mining for silver, gold, lead, copper, manganese, and zinc was already underway by 1869. In western Utah, ores of copper, lead, zinc, and gold were discovered almost simultaneously in the Oquirrh Mountains, where large-scale ore production at Bingham Canyon near Salt Lake City began in 1869. At almost the same time, mining activity commenced in the Tintic and San Francisco Mountains of western Utah. The enormous mineral wealth that incited this initial surge of prospecting and mining also brought geologists to many previously isolated and remote parts of the Great Basin. Geological studies of the ore deposits in the region revealed that the stratified rocks formed in primeval oceans were commonly involved in the mineralization process. Because they hosted some significant mineral resources, the origin and history of the ancient sediments were the focus of many geological studies for nearly a century.

As knowledge of the geological framework of the Great Basin expanded, it became clear that the oldest sedimentary rocks in the region could be traced beyond the area of internal drainage defined by Frémont. In the Mojave Desert of California and southern Nevada, a region within the Colorado River drainage system, rocks very similar in age and origin to those in the Great Basin proper were described and mapped in the twentieth century. The similarity between the oldest bedrock of the Great Basin and that of regions adjacent to it indicates that the internal drainage used by Frémont to define the region developed in the relatively recent geologic past. Geologists now have good evidence that the characteristic drainage of the Great Basin began to evolve no earlier than about 30 million years ago. Hundreds of millions of years before that development, the primordial seas must have uniformly covered much of what would eventually become western North America. Thus, the rocks that document the Great Basin seafloor, though characteristic of Great Basin mountains, extend beyond the modern drainage divide into northwest Arizona, southern Nevada, and southeast California. To acquire all the information necessary to reconstruct the captivating history of the ancient seafloor, we will also need to consider evidence from stratified rocks in these nearby areas.

Bedrock exposures in the Great Basin are most dramatic in the rugged, generally north-trending, mountain ranges that stagger across the region from the Wasatch Range in central Utah to the eastern escarpment of the Sierra Nevada. More than two hundred mountain ranges have been formally named in the Great Basin, and most of these were elevated along range-bounding faults in the relatively recent geologic past. The older rock that lifted as the mountains rose now forms the bold frontal escarpments and craggy walls of incised canyons. The arid climate and sparse vegetation of the Great Basin has limited both the weathering of rock outcrops and vegetative cover, leaving many rock exposures magnificently uncloaked—the tangible record of an unfathomably long history.

The geological evolution of the Great Basin, as we know it today, encompasses more than a billion years and reflects numerous transformations of the western margin of North America. We shall explore one of the most prominent episodes in the continuous geological remodeling of the western edge of our continent. This exploration is, of necessity, limited in scope. It would be impossible in a brief narrative to fully depict every interval in an inconceivably long geologic history, during which the Great Basin seafloor only prevailed during a portion of that time. How is this interval designated, and in what manner can it be subdivided?

The Geologic Time Scale

The current version of the geological time scale used by scientists to describe the history of our planet is the result of more than 400 years of development, beginning in the 1600s, as the modern science of geology arose from earlier traditions of natural philosophy and mysticism. The bewildering immensity of Earth time, which began some 4.6 billion years ago, is a perception that emerged gradually, as the fundamental concepts of geology were established and scientific technology advanced from the

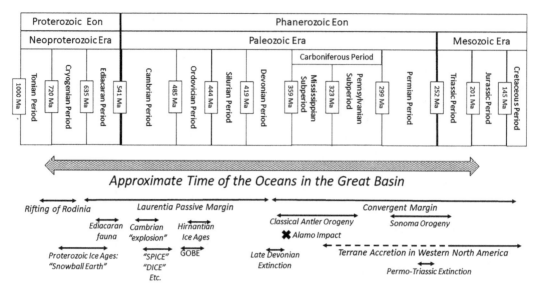

FIGURE 1.4. Timeline of the Great Basin seas and major geological and biological events that affected the region.

late 1700s. In fact, this process continues, even after several centuries of refinement. The time units summarized in Figure 1.4 are based on the 2018 update of the time scale compiled by the Geological Society of America (see Chapter 1 references for complete citation).

The fundamental unit of the geological time scale is the period. Each period was originally based on unique and characteristic assemblages of fossils, which were discovered to occur in layered rock sequences in a nonrandom pattern by William Smith in England between 1816 and 1819. Using Smith's *Principle of Faunal Succession*, geologists used distinctive sets of fossils to delineate discrete increments of prehistoric time on the assumption that any particular set of organisms coexisted for only a finite period of time. Though almost nothing was then known of the evolutionary processes or the mechanisms of extinction that resulted in the patterns, geologists in the early nineteenth century found that assemblages of fossils could be used consistently to identify periods of time in a sequence of stratified rocks. The various periods could then be assigned a relative age based on their position within the stratified succession, the older periods underlying younger periods. Each period was named in reference to the regions,

mostly in England and western Europe, in which the fossiliferous rocks were well exposed. Thus, the Devonian period was based on the set of fossils recovered from rocks near Devon (or Devonshire) County in southwest England. The earlier Cambrian period was named for rocks that yielded the definitive fossils in Cambria, a term applied to the United Kingdom's Wales region. The geographic significance of the period names may be more obvious to North American readers in the case of the Mississippian and Pennsylvanian subperiods, into which the European "Carboniferous" has been subdivided on this side of the Atlantic Ocean.

As more and more fossils were collected from widespread exposures of rocks representing various time periods, it became possible to group the periods into larger increments and to subdivide them into smaller segments, again using the patterns of fossil occurrence and ranges. Several periods can be grouped to comprise the larger eras, and similar eras can be assembled into even larger time units known as eons. There are four major eons in Earth history, the *Hadean* (4.6 to about 4.0 billion years ago), the *Archean* (4.0 to about 2.5 billion years ago), the *Proterozoic* (2.5 to about 0.541 billion years ago), and the *Phanerozoic* (541 million years ago to the present). The seas that submerged the modern Great Basin did so from the latest Proterozoic Eon to the Middle Phanerozoic, or from the Neoproterozoic Era to the early Mesozoic Era (Figure 1.4).

Subdivisions of the periods into smaller time units are based on sets of co-occurring fossils known as *zones*. There are several types of paleontological zones, but they all reflect the appearances and disappearances of specific fossils at precise positions in the layered rock successions. Using zones, periods of geological time can be divided into several epochs, which in turn can be further subdivided into smaller ages. For example, the Silurian Period is comprised of—from older to younger—the Llandovery, Wenlock, Ludlow, and Pridoli Epochs. The earliest of these epochs, the Llandovery incudes the Rhuddanian, Aerolian, and Telychian ages. As this hierarchical scale of geological time evolved over the past century, the once simple scale of relative time units became a morass of tongue-twisting names beyond anyone's ability to fully memorize.

In the latter half of the twentieth century, geologists began to develop several different techniques to establish the absolute ages of rocks based on the steady decay of radioactive elements, including specific isotopes of thorium, uranium, potassium, and others. While it is beyond the scope of our study to explore the details of these techniques, the actual beginnings and endings of the various subdivisions of the modern geological time scale are now established with reasonable, if not absolutely precise, accuracy. We now know, for example, that the Mississippian subperiod began about 359 million years ago (or *mega-annum*, abbreviated Ma) and ended about 323 Ma. Given the immensity of Earth history, the uncertainty in these relatively small determinations is negligible.

For the sake of simplicity, in our study of the prehistoric seas of the Great Basin, we will emphasize the various geological periods. To further reduce the potentially confusing tangle of jargon, we can utilize general terms such as the "Early Ordovician period" rather than specifying the Tremadocian and Floian Ages. The fascinating history of the Great Basin seafloor can be deeply explored and appreciated without memorizing all the formal subdivisions of the modern geological time scale.

Reading the Rocks: How to Reconstruct Vanished Oceans

The layers of rock exposed in the mountains of the Great Basin are pages of the Earth's autobiography. Each one records events and conditions of the past in the composition, texture, and structures preserved in solid stone that was once soft, seafloor sediment. In reconstructing the history of this seafloor, we will rely heavily on the *sedimentary* rocks that formed from what was initially gravel, sand, silt, muck, and mud. Over time, through a process known as lithification, the granular and chemical components of the squishy slurry were compacted and cemented into the rocks that constitute our portal into the seas of the Great Basin. What kinds of rocks are these, and how can we interpret them as historic documents?

Geologists traditionally divide the sedimentary rocks into two broad categories, partly overlapping, defined as detrital and chemical. The detrital sedimentary rocks are those that consist dominantly of granular materials, such as the familiar sandstone. The chemical rocks, on the other hand, are comprised mostly of chemicals precipitated from seawater as tiny crystals of minerals such as calcite (calcium carbonate, $CaCO_3$), quartz (silica, SiO_2), and halite (sodium chloride, $NaCl$). The chemical constituents of these minerals are always present, in varying concentrations, in seawater. Though this dichotomy of the sedimentary rocks is a useful concept, most sedimentary rocks consist of both detrital and chemical components, so the classification merely stresses which type of sediment is dominant.

Detrital sediment such as sand originates from the natural decomposition of rocks under the influence of the Earth's atmosphere, hydrosphere, and biosphere. To varying degrees, all rocks are susceptible to this process of decay, known to geologists as weathering. If you slide your hand over virtually any weathered rock

SEDIMENTARY ROCKS

Detrital
Conglomerate: lithified gravel; rounded grains generally larger than 2 mm. **Breccia** if rock fragments are angular in shape.
Sandstone: lithified sand; particles 1/16-2mm in size. **Orthoquartzite** (or just "quartzite" for some Great Basin sequences) if sand is nearly pure quartz; **Arkose** if sand has abundant feldspar grains with quartz; **Graywacke** if sand is "dirty" with numerous minerals and impurities.
Siltstone: comprised of "silt"; particles between 1/16 and 1/256 mm in size.
Shale: very fine grained rock comprised of "clay-sized" particles less than 1/256mm in size (microscopic); commonly thin-bedded or well-laminated.
Mudstone: a fine-grained rock consisting of a mixture of silt, and clay particles
Argillite: a rock similar to mudstone or shale, but more strongly lithified and with relatively weak laminations

Chemical
Evaporites: sedimentary rocks consisting of soluble minerals precipitated from water, such as halite (NaCl), gypsum (CaSO$_4\cdot$ H$_2$O), calcite (CaCO$_3$)
Chert: a hard and durable form of microcrystalline quartz (SiO$_2$) precipitated from seawater, and/or from siliceous organisms such as sponges or radiolaria.
Limestone: any sedimentary rock consisting dominantly of calcite (CaCO$_3$) precipitated directly from seawater or by organisms with calcareous skeletal material or shells such as corals, molluscs, brachiopods, and echinoderms. Limestone may be crystalline, fragmental ("bioclastic"), or both and is generally gray, tan, yellowish, or brown in color.
Dolomite or **Dolostone:** sedimentary rocks comprised mostly of the mineral dolomite, CaMg(CO$_3$)$_2$; dolomite is commonly light-colored (gray or tan) and crystalline, and forms in sedimentary environments similar to those for limestone.

FIGURE 1.5. The major types of sedimentary rocks.

surface, you are likely to see and feel the detrital products of weathering. These tiny fragments are the geological sawdust produced by the physical disintegration of bedrock. As such, they generally consist of the mineral grains that were firmly fused together in the unweathered parent rock. Such common minerals as quartz, feldspar, mica, and amphibole are common detrital grains.

The names applied to detrital sedimentary rocks emphasize the size, shape, and variety of the grains that comprise them. It surprises many people to learn that common terms such as sand have specific meaning to geologists. A particle of rock is only "sand" if it is between 1/16th and 2 millimeters (mm) in size. Silt grains are smaller, 1/16th to 1/256 mm in size, while pebbles are larger (4–64 mm).

The size ranges of the various particle names are specified by the Wentworth scale, first formulated in 1922, as follows:

Boulder: greater than 256 mm
Cobble: 64–256 mm
Pebble: 4–64 mm
Granule: 2–4 mm
Sand: 1/16–2 mm
Silt: 1/256–1/16 mm
Clay: less than 1/256 mm (microscopic)

Note that the Wentworth scale is not linear, and that the breadth of the categories is not always consistent. In more recent years, geologists have developed several other ways of describing particle sizes in detrital sedimentary rocks, but the basic names of the Wentworth scale are retained and carried forward in many of the rock terms. Siltstone and sandstone are the two best examples of such rocks. In addition to the size limits of each particle, geologists have also subdivided the categories into more specific size varieties, such as coarse sand (0.5–1 mm), as opposed to fine sand (0.25–0.125 mm).

Some of the names applied to detrital rocks signify a mixture of particle sizes and on occasion utilize other characteristics of the grains, such as shape or mineral composition. Mudstone, for example, is comprised of a mixture of mostly silt and clay, while conglomerate consists of lithified gravel, a mix of sand, granules, pebbles, and cobbles. Furthermore, some detrital rocks are subdivided into varieties based on sorting (purity and range of particle sizes). Graywacke, for example, is a "dirty" sandstone with several different minerals present as grains and some silt and clay between the sand particles (poor sorting). Orthoquartzite (also known as quartz sandstone) is a very "clean" sandstone, consisting of equidimensional sand grains (well sorted), composed dominantly of a single mineral, quartz. For future reference, the names of detrital sedimentary rocks important in the Great Basin are summarized in Figure 1.5.

Chemical sedimentary rocks consist mostly of mineral crystals precipitated from water. Aside from its obvious salt (present in seawater as sodium and chlorine ions), seawater also contains many other elements, among which carbon, calcium, magnesium, potassium, sulfur, and silicon are the most geologically important. The atoms of these elements often exist as chemically active ions (positively or negatively charged atoms or groups of atoms) that continuously react with each other to form a variety of solid compounds. These minerals can crystallize from seawater inorganically in the same way that sugar crystals grow from a water solution to make what is known as rock candy. The process of forming crystals from a solution is known as precipitation. Marine organisms also play major roles in precipitating minerals as material for shells, exoskeletons, and other durable tissues. As conditions of temperature, chemistry, and pressure change in the oceanic water column, the precipitated solids can dissolve back into ions. Whether they formed inorganically or through the activity of organisms, the solid minerals precipitated from seawater can accumulate on the seafloor without dissolving again. Such sediment consists of minerals formed in the oceans directly from seawater rather than the granular debris derived from weathering of rocks on land. After lithification, these soluble minerals comprise the bulk of chemical sedimentary rocks.

The chemical sedimentary rocks are generally subdivided based on composition; that is, which mineral is dominant among the several that might exist in the rock (Figure 1.5). Sometimes, chemical sedimentary rocks are given names reflective of specific events triggering precipitation. Evaporites, for example, are chemical sedimentary rocks that consist of minerals precipitated when evaporation raises the concentration of ions in the water to the point of saturation, triggering the growth of mineral crystals. Gypsum (calcium sulfate) and halite

(sodium chloride, or rock salt) are the two most common minerals in evaporite deposits.

Limestone is by far the most abundant chemical sedimentary rock forming in the shallow oceanic realm, both ancient and modern. It is an incredibly varied type of rock, but all forms consist almost entirely of calcium carbonate ($CaCO_3$) minerals, most commonly calcite or aragonite. Relatively pure calcite muds that accumulate on the seafloor typically form either visibly crystalline limestone or a dull variety known as micrite or "lithographic" limestone. However, limestone may also consist almost entirely of calcium carbonate formed by organisms such as molluscs (clams, snails, etc.), corals, echinoderms (star fish, urchins, and their relatives), and coralline algae. All these organisms use either calcite or aragonite in their skeletal material, protective shells, or tissues. Coquina is a form of limestone that consists almost entirely of shells and shell fragments composed of calcium carbonate. Limestone may also contain a mix of muddy sediment and shell fragments. Such varieties are commonly described as bioclastic limestone.

All types of limestone sediment form most effectively, and in greatest abundance, under specific ocean conditions. This is because calcium carbonate precipitates most readily from seawater that is relatively warm, shallow, clear, and well agitated by waves and currents. These conditions, which also favor the formation of calcareous shells by organisms, typically prevail today in shallow, tropical seas. In coral reef environments, where shell-secreting sea life is profuse, the production of carbonate sediment is extensive. The great volume of ancient limestone in the Great Basin, resulting from what some geologists have described as a "carbonate factory," suggests similar conditions in the past.

Dolomite is a calcium magnesium carbonate mineral, $MgCa(CO_3)_2$, that sometimes comprises the bulk of carbonate rocks. Many gelogists refer to such rock as dolostone, though in the Great Basin it has been historically more common for scientists to use the name dolomite for both the mineral and the sedimentary rock comprised of it. Dolomite, or dolostone, probably forms under the same general conditions as limestone, but the near absence of primary dolomite sediments in the modern world makes their precise origin a bit uncertain. We will discuss the "dolomite problem" in more detail later as we explore the Great Basin seafloor.

Chert is a hard and dense chemical sedimentary rock comprised of silica (SiO_2), primarily in the form of microcrystalline quartz. The silica in seawater comes from a variety of sources including rivers draining continents, groundwater discharge in coastal areas, submarine volcanic activity and hydrothermal vents, and wind-blown dust. Tiny crystals of quartz can be precipitated from seawater inorganically or through the activity of siliceous organisms such as sponges, diatoms, and radiolarian plankton. However, there is good geologic evidence that masses of chert can also form through replacement of other kinds of sediment, especially limestone and shale. Presumably, this process of replacement occurs during lithification but may begin even while the original sediment is still soft.

The origin of chert is not completely understood, but it generally occurs in two main contexts in Great Basin rocks. Bedded chert is sometimes found in association with dark-colored marine shale and siltstones deposited in relatively deep water. In today's oceans, such sediment generally accumulates as a siliceous ooze on the ocean floor under water more than about 15,000 feet deep. In this deep ocean environment, carbonate minerals tend to dissolve, while remains of siliceous plankton such as diatoms and radiolaria can remain intact as they drift to the deep ocean floor from the upper part of the water column. In addition to bedded chert, a nodular variety is commonly found as irregular lumps and masses in limestone and dolomite. This type of chert may form within the carbonate sediment as it accumulates in shallow marine environments, or it may form as replacement masses during or after lithification. Organisms such as sponges, which possess silica spicules and fibers, may also play an important role in the formation of chert nodules.

In reconstructing the depositional environment of sedimentary rocks, geologists also consider structures that can be preserved in the lithified sediment, whether it is detrital or chemical. For example, turbulent undersea landslides known as turbidity currents can result in the deposition of sediment that possesses graded bedding wherein larger particles at the base of the layer grade upward into finer-grained sediment (Figure 1.6A). Turbidity currents are generally triggered by earthquakes, submarine volcanic activity, or powerful storms that dislodge soft sediment that has accumulated on or near an undersea slope. Strong turbidity or other currents can also scour small, crescent-shaped depressions in the soft mud of the seafloor. Once sediment is deposited on the scoured surface, an inverse replica of the scoured hollow is preserved on the bottom of the overlying layer as a rounded mass known as a flute cast (Figure 1.6D). Both the scour marks, and the flute casts that replicate them can be used to determine the direction of current flow.

In addition, if the sediment accumulates under the influence of persistent directional currents, it commonly possesses one of several varieties of crossbedding in which fine, internal layers of sediments are oriented at an angle to the primary stratification (Figures 1.6B, 1.7A). Directional or oscillating currents can also lead to ripple marks in soft sediment that can be preserved after lithification as ridges and grooves along bedding planes. Both ripple marks and crossbedding can be used to determine the direction of sediment transport.

Soft sediment can also be disturbed prior to lithification to produce convolute bedding (Figure 1.6C). Convolute bedding may reflect semi-consolidated sediment sliding down a steep seafloor slope, seismic disturbances on the seafloor, a sudden influx of sediment, or dewatering that accompanies compaction of soft mud.

Finally, the activity of marine organisms crawling on, or burrowing through, the soft sediment can be preserved as trace fossils, or ichnofossils, in sedimentary rocks (Figures 1.6E, 1.7B). While these biogenic structures cannot always be associated with specific organisms,

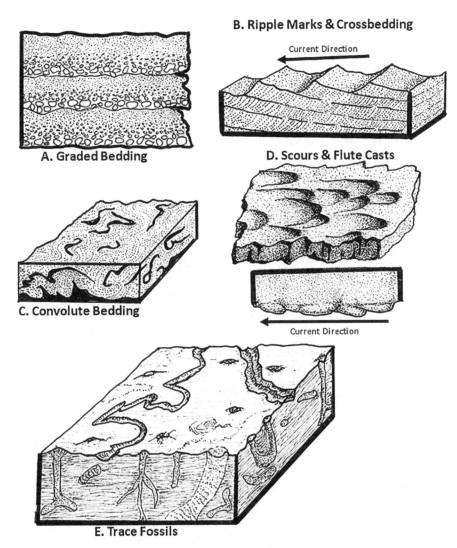

FIGURE 1.6. Common Sedimentary Structures: (A) graded bedding results from deposition of sediment via turbidity currents; (B) ripple marks and crossbedding generated by directional currents; (C) convolute bedding results from movement of unconsolidated sediments; (D) scours and flute casts generated by strong bottom currents; (E) trace fossils resulting from the crawling or burrowing activity of organisms.

the burrowing behavior they display is known to be influenced by such environmental factors as water depth, sediment grain size, wave and current agitation, turbidity, oxygenation, and distribution of food resources. All sedimentary structures are thus helpful in reconstructing depositional environments because they provide information not necessarily reflected by the sediment itself.

Stratigraphy:
The Study of Layered Rocks

Because sediment grains of all types accumulate under the influence of gravity, all sedimentary rocks originate as nearly horizontal layers of soft material. After lithification, the solid stone retains the layered form, and successive strata can build up thick sequences of layers. The thickness of each layer, or stratum (pl. *strata*),

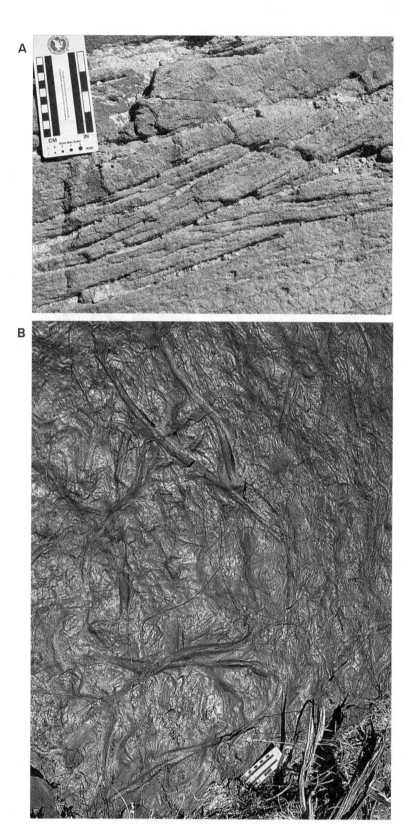

FIGURE 1.7. Crossbedding in sandstone of the Cambrian Tapeats Formation at Frenchman Mountain, Nevada (1.7A), and trace fossils in Cambrian shale of the Campito Formation, White-Inyo Mountains, eastern California (1.7B). Smallest divisions on scale bars in 1.7B are centimeters.

is influenced by the grain size of the sediment, rate of sediment accumulation, the periodicity of deposition, degree of compaction during lithification, and even some postdeposition factors. An individual stratum may be a millimeter thick in laminated shale or several meters thick in more massive limestone.

Stratigraphy is the branch of geology devoted to the study of layered sequences of rocks. One of the fundamental ideas in stratigraphy is the concept of a formation. A formation is a sequence of rock layers that have similar characteristics such as grain size, composition, bedding thickness, and sedimentary structures. The combination of such features serves to distinguish the rocks in a particular formation from other layered sequences above and below it. Since the rocks that comprise a formation are more or less uniform in their characteristics, they represent sediment that accumulated under similar conditions in similar environments. The boundaries between successive formations are placed where the characteristics of the rocks change, for example where coarse sandstone is overlain by a sequence of fine mudstone layers. Such shifts represent changes in sediment accumulation that, in turn, suggest changes in the environment of deposition. Formations, then, represent chapters or phases of generally constant conditions in the overall continuum of environmental change recorded by successive formations.

Formations are generally named for the localities where the rock unit is particularly well exposed or was first studied and defined. The Roberts Mountains, for example, is a prominent mountain system in Eureka County, Nevada, about 30 miles northwest of the town of Eureka. The Roberts Mountains Formation consists of sedimentary rocks that are best exposed in that location, even though the layers extend beyond the mountains themselves and have been identified in many other parts of the Great Basin. The formation consists of limestone, shale, and dolomite of fairly uniform characteristics. On occasion, when a particular type of rock comprises nearly all of a formation, the word "formation" is replaced with an epithet reflective of that dominant lithology, or rock type. This is the case in formations such as the Eureka Quartzite and the Pioche Shale. The thick and varied Paleozoic rock sequence in the Great Basin has been divided by geologists into hundreds of different formations, most with type localities in or near Great Basin mountains.

In many cases, it is possible to subdivide a formation into smaller sequences or intervals that represent a variation of the overall geological characteristics. Such subdivisions of formations are called members. The Pioche Shale of eastern Nevada has been split into several members in most of its eastern Nevada and western Utah exposures. Each of these members is given a locality-based name in the same manner as the formation to which they belong. Thus, in Lincoln County, Nevada, the Pioche Shale consists of, in ascending order, the Delamar Shale Member, the Combined Metals Member, the Comet Shale Member, the Susan Duster Limestone Member, the Log Cabin Member, and the Grassy Spring Member. All these subdivisions of the Pioche Shale represent variations on the shale-dominated strata that comprise the overall formation. In a similar manner, it is sometimes feasible to combine several formations into larger entities known as groups. A group is a set of formations that have some overall similarity, even though they differ individually. For example, the Pogonip Group in the Monitor Range area of the central Great Basin is more than 1,100 feet thick and consists of, in ascending order, the Goodwin Limestone, Ninemile Shale, and the Antelope Valley Limestone. All three of these Ordovician formations are dominated by limestone and shale, but the details of their texture, composition, sedimentary structures, and other features are individually distinctive. Thus, although all three formations consist principally of shallow oceanic sediments, different kinds of carbonate and granular sediment were deposited under slightly different conditions in the time recorded by the Pogonip Group.

The rocks that document the Great Basin seafloor consist mostly of sediment deposited in a variety of marine environments. At various times in the immense period of oceanic inundation, sediment accumulated on tidal flats, around reefs constructed by corals and other organisms, on sandy beaches and offshore sand-bars, on steeply sloping continental ramps, on shallow, undulating shoals, and in deep, murky basins. As in modern oceans, all these marine settings had finite extent on the Great Basin sea-floor. Sandy beaches may have passed offshore into coral reefs, which in turn may have bor-dered a deep ocean basin. Under such circum-stances, different types of sediment would have accumulated at the same time in adjacent areas of the ocean floor. Therefore, it is not surprising to learn that many Great Basin rock formations pass laterally from one type of rock into another. In the Tintic Mountains of western Utah, the Tintic Quartzite consists of sandy materials that accumulated on beaches and in offshore sandbars during the Early Cambrian Period. At approximately the same time, fine-grained mud accumulated in deeper water farther off-shore in what is now western Utah and eastern Nevada. These sediments comprise much of the Pioche Shale, a formation distinct from the Tin-tic Quartzite but approximately the same age. Because conditions on the seafloor varied from place to place and over time, we will see many such examples of time-equivalence of various formations across the region. In addition, the sedimentary rocks are sometimes associated with volcanic materials that represent under-sea eruptions or the drift of volcanic ash over the ancient oceans. Deciphering the conditions under which the sedimentary rocks in the Great Basin originated involves assessing the clues provided by the sediment type, sedimentary structures, fossils, and certain chemical charac-teristics of the rocks.

As we begin to explore the rock record of the Great Basin seafloor and piece together its extensive history, we will do so based on what the rocks and fossils tell us about the oceanic conditions under which they formed hundreds of millions of years ago. And, as you may have already guessed, they tell us a lot!

Plate Tectonics and Oceans

Contrary to what many people assume about oceans and continents, neither are permanent features of the Earth's surface. The slow move-ments of the rigid lithospheric plates that com-prise the outer 100 km of the planet result in a constant reconfiguration of land and ocean basins, rather like a slow-motion kaleidoscope. As we will learn in the coming chapters, the Great Basin seafloor developed beneath an ocean and adjacent to a continent very differ-ent from its modern counterpart. Before we explore the rock record of nearly a billion years of inundation in the Great Basin, it will be help-ful to reflect briefly on how ocean basins and the surrounding continents are shaped by plate tectonic interactions.

The lithosphere of the Earth consists of great plates of rock that rest on a roughly 100 km deep layer known as the asthenosphere (Figure 1.8). The rock that comprises the asthenosphere is unusual. Although no one has ever seen this material directly, geologists have considerable indirect evidence that the material comprising the asthenosphere is like peridotite, a dense (3.3–3.4 g/cm^3) igneous rock composed pri-marily of iron and magnesium silicate minerals. However, under the conditions of temperature and pressure that prevail in the deep earth, the peridotite is less rigid than at the surface and can flow under the influence of heat, pressure, and gravity. The idea of a solid rock behaving as a soft, mushy material is a bit contrary to our everyday experience at the surface, but the movement of seismic waves though the astheno-sphere clearly indicates its ductile nature. Thus, the rigid plates of the lithosphere are not held firmly by the softer, but generally denser, asthen-osphere below. Without an underlying anchor, the lithospheric plates can move if forces from gravity or heat-generated movement of the asthenosphere are applied to them. This is the essence of the global plate tectonic system within the lithosphere.

Integrated studies of the lithospheric plates utilizing seismic methods, gravity data, and deep drilling reveal that the plates are con-structed of two components (Figure 1.9). The lower portion of the plates is comprised of dense and rigid rock similar in composition and density to the softer peridotite of the underlying asthenosphere. Above the dense and rigid base of the lithosphere, across a boundary known as the Moho, a layer of less dense rock forms

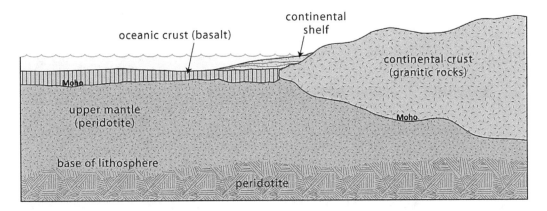

FIGURE 1.8. The lithosphere of the earth is about 100 km thick and is comprised of upper mantle peridotite and the overlying crust. The Moho is the boundary between the crust and mantle within the lithosphere. From DeCourten and Biggar (2017), used by permission of the authors.

the crust, or outer skin, of the lithosphere. The crustal material of the plates is of fundamentally different nature under the oceans than on land. These differences are important in understanding the plate tectonic interactions involved in the opening and closing of ocean basins through time.

The upper 5–10 km of oceanic lithosphere consists of igneous rocks formed by undersea eruptions and subsurface emplacement of lava. Overall, the rock of the upper oceanic lithosphere is like basalt, a common volcanic rock with a density near 3.0 g/cm³. Under land areas, the continental lithosphere possesses a thicker upper crust (up to about 60 km) of rock like granite but that contains far less iron and magnesium than basalt and is less dense (~2.7 g/cm³). The fact that most oceanic lithosphere is submerged while continental lithosphere is exposed as dry land reflects the difference in the thickness and density of the crustal material the plates carry. Relative to the thinner and denser oceanic plates, continental lithosphere is thick and buoyant, rising several kilometers higher than the surface of the basaltic seafloor rocks. The floor of the modern ocean is therefore submerged *because* it is underlain by low-lying oceanic lithosphere.

There are some places in both the modern and ancient world that the low margins of continental plates are, or were, submerged as well. These marginal seas are typically referred to as epicontinental seas. However, most of the water in the oceans exists where the lithospheric plates carry oceanic crust and form vast, low basins on a global scale. Before a seafloor (in the Great Basin or elsewhere) can exist, oceanic lithosphere must form. This process involves divergent boundaries between plates, where each plate moves away from each other. It is in the gap created along the divergent plate boundary that new oceanic lithosphere forms.

In the modern Earth, most oceanic lithosphere originates at the oceanic ridge system that wraps around it like stitches on a baseball. Along this almost entirely submerged mountain chain, the process of seafloor spreading continuously creates new oceanic lithosphere where two plates pull apart and move in opposite directions. This plate motion stimulates the production and movement of molten rock (magma) into the fractures and gaps in and between the overlying plates. Eventually, as the plates drift apart at a few centimeters per year, the leading edges sink back into the deep earth via the process of subduction (Figure 1.10). Subducting plates are ultimately destroyed, with the material recycled in the deeper portion of the mantle. Worldwide, there is rough balance between the amount of new oceanic lithosphere created at the oceanic ridges and the amount of older lithosphere consumed in subduction zones. In individual ocean basins, however, there can be more lithosphere created than

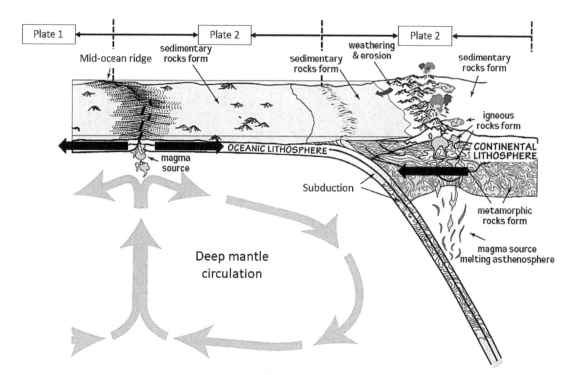

FIGURE 1.9. Plate tectonic system of the Earth. Plates form at divergent boundaries and are consumed by the process of subduction at convergent boundaries. The oceanic lithosphere is generally submerged because the crust it carries is thinner and denser than continental lithosphere. Modified from a USGS graphic.

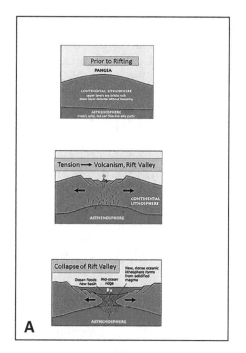

A

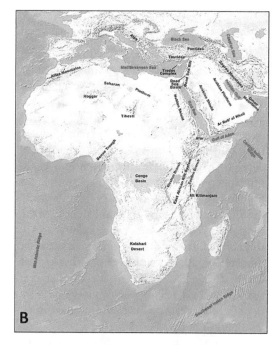

B

FIGURE 1.10. (A) The rifting of continental lithosphere results in the opening and progressive widening of a new ocean basin and ridge system. (B) Geologically recent rifting of the African plate has separated the Arabian plate from the larger continent via the opening of the Red Sea and the Gulf of Aden. A: modified from a USGS graphic; B: from USGS Open-File Report 2005-1994-E.

Eustatic Sea Level Changes

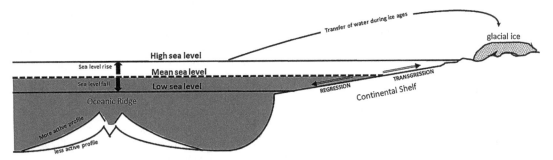

FIGURE 1.11. Eustatic sea level changes. Sea level rises and falls through time primarily in response to climatic, plate tectonic, or sedimentation factors that influence the capacity of ocean basins and the volume of water they contain.

destroyed, and the ocean basin will grow larger over time. This has been the case in the Atlantic Ocean basin for the past 180 million years or so. If the rate of subduction is higher than the rate of seafloor spreading, then the ocean basin will tend to shrink over time. Because these plate tectonic processes are continuous, no ocean basin or adjacent continental mass is static. Ocean basins open and close, and the continents they separate move about, in a relentless reconfiguration of the Earth's lithospheric plates. Nothing lasts forever on the dynamic Earth.

A variation on the theme of plate divergence that will be important in the story of the Great Basin seafloor is the process of continental rifting, wherein a single plate of continental lithosphere is separated into two or more segments that move away from a nascent oceanic ridge (Figure 1.10A). Northeast Africa, where the Arabian Peninsula has been severed from the African plate, is the best example of continental rifting in the modern world (Figure 1.10B). Both the Red Sea and the Gulf of Aden are very young ocean basins that opened initially only a few million years ago. Young oceanic ridges are developing on the floor of each of the narrow basins. The process of continental rifting not only creates a new ocean basin, but it also forms two new plates, each carrying a riven fragment of continental lithosphere in opposite direction. As the landmasses move steadily apart, the ocean basin also becomes wider.

Fitful Seas

Once an ocean basin forms, the amount of water that can accumulate in it changes continuously as a function of several factors. Anything that changes the volume of water or the capacity of the basin to store it will result in a variation in the size and extent of the seas over time. Sea "level," it turns out, is not so level after all. Obviously, the rate of seafloor spreading is an important consideration in the storage capacity of a new ocean basin. The higher this rate, the faster the ocean basin widens, and the more water it can accommodate. However, other circumstances are also important in the long-term evolution of an ocean basin.

Global climate is one of the factors that has an important influence on sea level. For example, during ice ages, water is transferred from the oceans to land as glacial ice much faster than it returns as melt water. Sea level therefore falls during glacial times. Twenty thousand years ago, at the peak of the most recent ice age, sea level was approximately 130 meters (400 feet) lower than it is today. Conversely, warm intervals that cause rapid melting of glacial ice result in a sea-level rise, a situation that is clearly in progress today because of rapid global warming. Climate-induced changes in sea level result in either the advance of the oceans inland, as sea level rises, or their withdrawal from the continental margins, as sea level falls. The advances are known as transgressions, while retreating seas result in marine regressions. Today, we live in an era of geologically rapid transgression, due primarily to the escalating warming of the global climate through human activity. In the modern world, low-lying coastal areas currently face serious threats from the global transgression that began 100 years ago and is accelerating at a frightening pace.

Another factor that influences sea-level over time is the rate of seafloor spreading that occurs at the oceanic ridges. In the modern world, this rate varies from about 2 cm/year along the oceanic ridge in the Atlantic Basin to nearly 18 cm/year in the eastern Pacific Ocean Basin. Studies of the volumes and ages of ancient oceanic lithosphere clearly demonstrate that the rate of seafloor spreading varies over time as well. Recall that new oceanic lithosphere is formed from the emplacement and eruption of magma along the axis of the oceanic ridges. When the rate of seafloor spreading is high, great volumes of hot magma rise into a ridge, causing it to expand into a broader and higher edifice. In contrast, if the rate of seafloor spreading is low, less magma is injected into the ridge axis, and it will therefore contract into a relatively narrow and low structure. In this manner, oceanic ridges can swell and shrink over time as they become more active or less active. Any change in the volume of the oceanic ridges will influence the capacity of the basin to store seawater. Bloated, active ridges will displace water landward, causing transgression and rising sea levels. On the other hand, when an oceanic ridge becomes less active, sea level falls and regression can result.

In addition to global climate and plate tectonic influences, sea level is affected by other, generally less significant factors that are, in fact, linked to the combination of climate and tectonic events. For example, the rate of sediment accumulation and basin subsidence can influence sea level because either may reduce or increase the storage capacity of an ocean basin. However, the rate of sediment accumulation is, in part, a consequence of the rate of weathering on land, which is in turn strongly influenced by climate and plate tectonic events such as mountain-building. Also, subsidence of the seafloor can occur as plate motion carries hot and young oceanic lithosphere laterally away from the ridge axis where it was created by seafloor spreading. As the young and hot oceanic lithosphere ages and cools, it contracts to become denser and settles deeper into the asthenosphere, carrying the seafloor downward with it. Subsidence of the ocean floor can also be influenced by the thickness and rigidity of the lithosphere, both of which reflect the tectonic history of the plate.

Global changes in sea level resulting from any combination of the multiple factors that influence water volume and oceanic storage capacity are known as eustatic sea-level variations. All eustatic changes result in global transgression or regression, the extent of which depend on the magnitude of the change in sea level. As we will see, conditions across the Great Basin seafloor were strongly influenced by numerous eustatic variations spanning hundreds of millions of years. Many of the transgressions and regressions were minor, but several can be correlated with major excursions of sea level that affected the entire planet.

Our toolbox of basic concepts used to construct visions of past worlds from the geological evidence preserved in rocks is now complete. Plate tectonics, an acknowledgment of the immensity of geologic time, the realization that sedimentary rocks can be used to document ancient events and conditions, and an understanding that oceans are never static through time are all that is really needed to explore in detail the captivating history of the Great Basin seafloor. We can now begin our adventure.

2

The Ocean Arrives

Through the perpetually changing arrangement of lithospheric plates on the Earth, global geography is subject to dynamic transformation through time. The configuration of modern ocean basins and continents has only existed for about the last 4 percent of earth history. Earlier patterns of land and sea bear little resemblance to a current world map. As continents move with the plates that bear them, adjacent ocean basins shrink or expand—even opening and closing entirely—in an endless tectonic tango. The story of the Great Basin seafloor begins in an era of global geography profoundly different from that of today. Near the beginning of the Neoproterozoic era, about 1 billion years ago, the continental lithosphere was almost entirely melded into a single enormous edifice: the supercontinent of Rodinia (Figure 2.2A).

Origin of Rodinia

Although it has long since been ripped asunder by plate tectonic processes, geologists have reconstructed ancient Rodinia through analysis of the nature, distribution, and magnetic properties of the oldest rocks comprising the cores of the modern continents. In North America for example, the Neoproterozoic and older rocks constitute the original core, or shield area, centered in the Great Lakes region (Figure 2.1). Detailed mapping and dating of these ancient rocks reveal several age provinces that reflect the long, step-by-step construction of our continent's primordial ancestor. Each of the pre-Neoproterozoic Age provinces represents a mass of rock that became sutured into the ancient core over a long history of convergent plate interactions. This ancient core of North America became the central portion of Rodinia, with the shield areas of other modern continents tectonically fused to its margins (Figure 2.2). The heart of Rodinia is known as Laurentia, the part that eventually became the geological nucleus of modern North America. The formation of Rodinia was a gradual process that required more than 400 million years to complete. The cores of the modern continents, themselves amalgamated by earlier collisions, merged to form the gigantic landmass one piece at a time. By about 1 billion years ago, the process of assembling Rodinia was complete, and the gigantic supercontinent was positioned along the ancient equator and surrounded by a single global ocean.

The Great Basin region at the time of Rodinia was a vast undulating terrain situated in the interior of the supercontinent. Erosion was the dominant process, so only limited amounts of land-deposited sediment accumulated in local basins across the region during most of the Neoproterozoic era. The landscape and environment of Rodinia would have seemed like an alien world. The atmosphere contained only about 1 percent of the oxygen it has currently. Without much oxygen, only tiny traces of stratospheric ozone—our primary shield from harmful solar radiation today—could have formed.

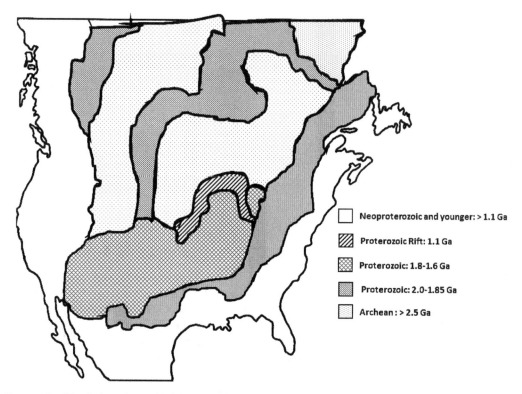

FIGURE 2.1. North American shield, essentially equivalent to the ancient continent of Laurentia, was comprised of several age provinces. Each of the patterned areas represents a mass of rock accreted to Laurentia prior to the Neoproterozoic Era, about 1 billion years ago.

Legend for Figure 2.1:
- Neoproterozoic and younger: > 1.1 Ga
- Proterozoic Rift: 1.1 Ga
- Proterozoic: 1.8–1.6 Ga
- Proterozoic: 2.0–1.85 Ga
- Archean : > 2.5 Ga

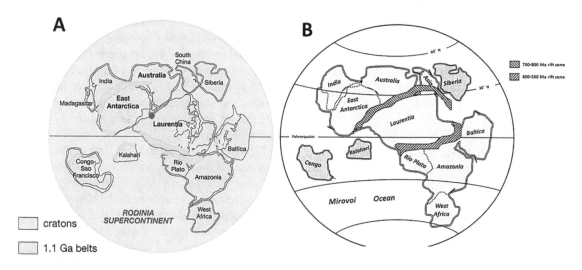

cratons
1.1 Ga belts

FIGURE 2.2. (A) The supercontinent Rodinia at the time of its final assembly about 1 billion years ago. Note that Laurentia, the core of modern North America occupied a central location in the supercontinent. The red dot represents the approximate location of the modern Great Basin. (B) Two major rift zones developed during the early stages of the rifting of Rodinia. In the Great Basin region, the rift zone was most active 700–800 million years ago. A: modified from DeCourten and Biggar (2017), used by permission of authors.

Consequently, the land surface was essentially sterilized by the blistering ultraviolet radiation of the sun. There were no land plants during the Neoproterozoic, so the surface was a barren, rocky wasteland. Surrounded by a single large ocean, the climate of Rodinia would have been extreme, with alternating scorching droughts and rampaging floods and, as we will soon learn, frigid ice ages. When the rains came to the interior of Rodinia, the raging flood waters were unrestrained by vegetation and scoured great channels into the bedrock while washing the rocky rubble into the lower basins.

The vast global ocean surrounding Rodinia is known to geologists as the Mirovoi Ocean, a name derived from the Russian term for "world" or "global." Conditions in the Mirovoi Ocean were also profoundly dissimilar to our modern seas. The salinity of Mirovoi seawater was probably close to that of modern seawater. However, the deeper waters were nearly devoid of oxygen, while the vigorous weathering and erosion on land increased seaward flux of such metallic elements as iron and nonmetals such as sulfur. The organisms that facilitate the modern nitrogen and phosphorous cycles had either yet to evolve or were extremely rare. Consequently, phosphorous and nitrogen, both important as nutrients for modern marine organisms, were limited in the Neoproterozoic seawater. The unusual chemistry and strong stratification of the Mirovoi Ocean restrained life to a fauna of mostly primitive, unicellular, photosynthetic plankton that could survive in the near-surface oxygenated water. These organisms may have been like modern cyanobacteria; they appear to have lacked diversity in both form and function and evolved very little during the time of Rodinia.

The Breakup of Rodinia

Rodinia remained intact for about 150 million years after its assembly. Then, in the Middle Neoproterozoic Era, starting around 825 million years ago, continental rifting began to disassemble the supercontinent. The period of rifting lasted more than 100 million years, during which several portions of the supercontinent were broken apart in different places and at different times. In what is now the Great Basin, one of several major rift zones developed incrementally between about 800 and 725 million years ago. This large rift zone ultimately separated the west side of Laurentia, the core of modern North America, from the rest of Rodinia. To the west (by today's geography), across the rift zone from the margin of Laurentia, several fragments of Rodinia were pulled away and are now embedded in the cores of modern Antarctica and Australia (Figure 2.3). As with eastern Africa today, the great rift zone in western Rodinia eventually sank below sea level, its fragments moving in opposing directions. Ultimately, a narrow, nascent ocean basin formed as water from the Mirovoi Ocean submerged the sinking floor of the rift. The Great Basin seafloor was born.

The rifting of Rodinia occurred in response to heat build-up beneath the supercontinent over hundreds of millions of years. Heat from the searing core of the young Earth mobilized the softer rocks of the asthenosphere and expanded the overlying continental lithosphere into a dome-like structure. Tensional forces began to stretch the continental crust to the breaking point, resulting in the collapse of the dome along linear, rift-valley faults. The separation of unfractured rock on either side of the rift valley reduced the weight pushing down on the overheated mushy rock of the underlying asthenosphere. The lateral displacement of overlying rock caused a simultaneous reduction in the melting temperature of the asthenosphere. This caused great volumes of magma (molten rock) to form in the subsurface via a process that geologists call decompression melting. Driven upward by internal pressure, the magma flowed onto the floor of the deepening rift valley in vast lava flows. Some of the molten magma hardened in the subsurface to form rocks such as the peridotite of the deeper lithosphere. Once the floor of the valley was submerged by seawater, continuing seafloor eruptions began constructing an oceanic ridge. Seafloor spreading away from the axis of the newborn ridge became active, and the fragments of old Rodinia were carried farther apart as the seaway widened.

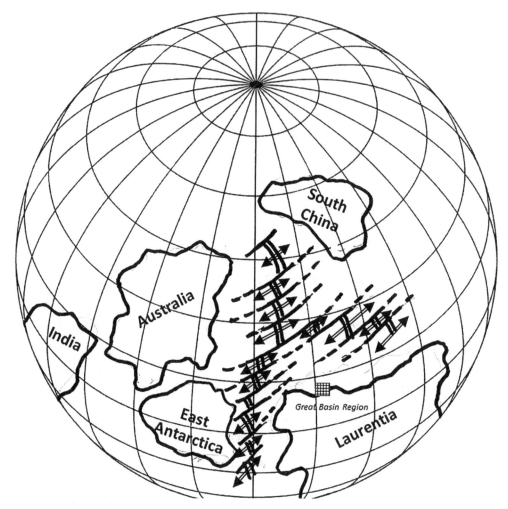

FIGURE 2.3. Fragments of Rodinia west of Laurentia now comprise portions of the interiors of modern Asia, Antarctica, and Australia.

The new oceanic lithosphere created by this process was the initial bed of the Great Basin seafloor.

By about 700 million years ago, several fragments of Rodinia had become isolated from the supercontinent and were attached to new plates that formed during the rifting process. One of those newly created plates carried Laurentia, the fragment of central Rodinia now separated as a discrete continent. The narrow but expanding ocean basin between Laurentia and other fragments of former Rodinia would eventually grow into a global body of water referred to as the Panthalassa Ocean. The Great Basin region was situated along the rifted northern margin

of Laurentia and very near the coast where the barren land met the waves of the Panthalassa Ocean, the successor to the Mirovoi Ocean.

Neoproterozoic Ice Ages

As Rodinia was breaking apart during the Neoproterozoic, the Earth slipped into what is probably the most severe series of ice ages in the long history of our planet. So widespread were these multiple cold intervals, that once geologists recognized their global extent, the term Cryogenian was established for the middle period of the Neoproterozoic Era. The Cryogenian Period spans the interval from about 720 to 635 million years ago. Note that this period coincides with the later stages of the isolation of Laurentia from Rodinia and the initial opening of the Panthalassa Ocean. Even as Rodinia broke up, vast sheets of ice covered much of its frigid surface and some parts of the ocean.

There were at least four periods of nearly world-wide glaciation during the Neoproterozoic. The two most significant of these occurred during the Cryogenian Period: the Sturtian glaciation (thought to have occurred from about 715–700 million years ago) and the later Marinoan glaciation (660–630 million years ago). These two global ice ages were separated by a brief interglacial episode and were preceded and followed by less significant cold intervals. Collectively, the two main Cryogenian glaciations have inspired the term "Snowball Earth" as an epithet for the time of extreme glaciations. Judging from the sedimentary evidence and carbon isotope record preserved in Cryogenian deposits, Snowball Earth conditions were truly remarkable. At the apexes of the cold conditions, land was covered with ice more than a kilometer thick, while most of the ocean was also ice-capped, and even tropical seas were slushy with ice. Global mean temperatures may have been as low as $-12°C$ ($10°F$), and conditions along the equator of Snowball Earth may have been not much different from the interior of modern Greenland.

The causes of the Cryogenian ice ages are still being debated by geologists and climatologists. As is true of most climatic excursions, there were probably multiple factors that drove the planet into the episodic deep freezes. The Earth's climate system was undoubtedly different in the Cryogenian Period than it is now. The days were a bit shorter; the sun was about 6 percent less luminous; the atmosphere contained only small amounts of oxygen; life was not diverse or complex; and the moon was closer. These differences make it difficult to model how orbital cycles might have affected global climate. However, many geologists are convinced that the formation and break-up of Rodinia, through their effect on global geography and environment, were significant factors in the Cryogenian ice ages. Before and after the rifting, most of the land area on the Earth was positioned near the equator during the Early Cryogenian Period (Figure 2.2). Under the relatively warm and moist preglacial conditions along the equator, the weathering of silicate minerals in the ancient rocks would have been intense. This process consumes carbon dioxide from the atmosphere, reducing its ability to absorb warmth radiated from the surface. The Cryogenian world may have experienced "reverse" greenhouse conditions, where declining atmospheric CO_2 cooled the global climate.

In addition, the smaller landmasses from the Rodinian fragmentation would also have affected climate and weathering rates by allowing more moisture to reach rocks exposed in previously arid interior regions. Rocks weather more intensely in moist climates than in dry conditions, so the moderation of global aridity would have enhanced the drawdown of atmospheric CO_2. These cooling influences on climate were probably very subtle initially, but as more of the world became covered with highly reflective ice, less solar radiation warmed the surface. The expansion of ice is a positive feedback mechanism that would accelerate and intensify an initial cooling trend to the point that a full-fledged ice age enveloped much of the planet.

On the other hand, as the rifting of Rodinia progressed, some of the separating continents began to move away from the equator, and rapid seafloor spreading widened the narrow seaways. This might begin a trend toward less intense weathering of interior rocks, lowering the rate of withdrawal of atmospheric CO_2. Volcanic activity in developing rift valleys and at the axis of the nascent oceanic ridges became intense, with great volumes of CO_2 emanating from the erupting lava. These factors could have led to significant increases in atmospheric CO_2 and other greenhouse gases, reversing earlier trends toward cooler conditions. Though there are still many uncertainties about the fundamental causes of Cryogenian ice ages, it is very likely that the tectonic history of Rodinia was a critical factor in the pulses of glaciation that caused Snowball Earth during the Neoproterozoic Era. Geological evidence of glaciation clearly suggests that the factors driving the Earth to a

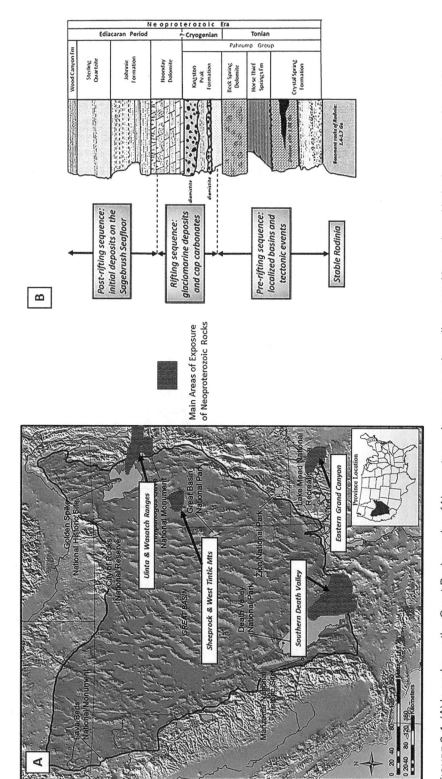

FIGURE 2.4. (A) In and near the Great Basin, rocks of Neoproterozoic age are only well exposed in northern Utah, the eastern Grand Canyon, and the southern Death Valley region. (B) The Pahrump Group of southern Death Valley records the rifting of Rodinia and the birth of the Great Basin seafloor. Base Map in A from U.S. National Park Service.

colder condition were stronger than those that tended to warm the planet during the Cryogenian Period.

The Geological Record of Neoproterozoic Rifting and Glaciation in the Great Basin

Rocks that record the events of the Middle and Late Neoproterozoic Era are not widespread in the Great Basin region, but there are four main areas where outcrops of strata that formed during the time of Rodinia rifting and contemporaneous glaciations are reasonably well exposed (Figure 2.4A). In the Uinta and Wasatch Ranges of northern Utah, thick sequences of sediment that comprise the Uinta Mountain and Big Cottonwood Groups accumulated during the Middle and Late Neoproterozoic. To the southwest, similar rocks comprise the several formations of the Sheeprock Group, exposed in the immediate vicinity of the Sheeprock and west Tintic Mountains. In the eastern Grand Canyon, near but well east of the Great Basin, the Chuar Group consists primarily of sandstone, shale, dolomite, and conglomerate strata more than a mile thick that accumulated during the initial break up of Rodinia. However, the best record of Neoproterozoic events within the Great Basin are the sediments of the Pahrump Group, widely exposed in the southern Death Valley region of southeast California and southern Nevada. These rocks provide some intriguing insights into the inception of the Great Basin seafloor.

The Pahrump Group of the Southwestern Great Basin

The Pahrump group consists of several formations that were deposited during the time that Rodinia was breaking apart and the world experienced the Snowball Earth glaciations. Originally named in 1940 for exposures near the Pahrump Valley of southern Nevada, the Pahrump Group consists of more than 13,000 feet of sedimentary rocks separated into four formations that span about 450 million years of the Neoproterozoic time interval (Figure 2.4b). The oldest rocks in the Pahrump Group are approximately 1.1 billion years old, and the

youngest sediments accumulated about 650 million years ago. Thus, the oldest strata of this thick rock sequence were deposited just before the rifting of Rodinia began, and sediment continued to accumulate through the Cryogenian Period as the global climate deteriorated and the earliest ancient ocean basins began to develop in the Great Basin region. Geologists have identified several major gaps, known as unconformities, between and within the formations of the Pahrump Group, so the geological record of the inception of the Great Basin seafloor is not complete or continuous. Nonetheless, the sediments of the Pahrump group allow scientists to reconstruct the general sequence of events through which the seafloor originated.

The lowermost and oldest formation of the Pahrump Group is the Crystal Spring Formation, consisting of a mixed assemblage of sandstone, conglomerate, dolomite, and diabase, an igneous rock representing ancient lava. The Crystal Spring sediments rest on the eroded surface of pre-Neoproterozoic metamorphic and igneous rocks (1.4–1.7 billion years old) that represent the geological foundation of Rodinia prior to its rifting. The lowermost strata of the Crystal Spring Formation are mostly coarse, feldspar-rich sandstone and conglomerate that appear to have been deposited by swift streams carrying sand and gravel from elevated portions of Rodinia north of the Death Valley region into an interior basin. This ancient basin may have subsided along faults that signify the beginning of the collapse of the rift valley that ultimately severed the Rodinia landmass. The coarse, lithified rubble at the base of the Crystal Spring formation grades upward into finer-grained sandstone and mudstone that appears to have accumulated in a tidal flat and/or delta setting. Such sediments record the initial arrival of marginal marine conditions in the southwestern Great Basin.

The upper portion of the Crystal Spring Formation is dominated by laminated limestone and dolomite, sedimentary rocks that clearly document shallow marine conditions. These carbonate rocks were probably deposited in a small embayment, rather than a vast seafloor, because they grade southward into mudstones and siltstones that consist of fine-grained sediment derived from land exposed a short distance south of the Death Valley region. In the Kingston Range of southeast California, the carbonate strata of the Crystal Spring Formation are sometimes cut and intruded by lava bearing a radiometric age of about a billion years. This lava was injected into the sediments after they had lithified, so the age represents the minimum age of the enclosing limestone and dolomite. Metamorphism of the Crystal Spring carbonate rocks by the invading lava has produced significant mineral deposits, including talc and asbestos, in at least 45 different mining areas in the southern Death Valley region.

Overlying the Crystal Spring Formation in the Pahrump Group is the Horse Thief Springs Formation, named for a locality in the Kingston Range south of Death Valley. A prominent unconformity at the base of that formation, separating it from the underlying Crystal Springs Formation, signifies an extensive interval of erosion prior to the deposition of sediments atop the underlying shallow marine carbonate strata. This is the first of several major gaps in the rock record of the Pahrump Group and probably symbolizes more than 300 million years of unrecorded time. This lengthy gap probably resulted from renewed uplift that expelled the restricted seas from the southwest Great Basin and temporarily exposed previously deposited sediments to erosion. It appears that the initial incursion of seawater into the region was reversed for a time by tectonic and perhaps volcanic activity adjacent to the nascent rift valley. When sediment deposition resumed, more than 2,000 feet of sandstone and mudstone associated with thin dolomite sequences accumulated as the Horse Thief Spring Formation. The sediments seem to have accumulated along a sandy beach, in tidal basins, and on a shallow, subtidal seafloor in several cycles. Thus, the rocks of the Horse Thief Springs Formation indicate the return of marine or marginal marine conditions due to the renewed subsidence of the rift valley.

However, the seafloor and beach on which the sediments accumulated were neither deep nor extensive. Studies of zircons deposited in the conglomerate at the base of the Horse Thief Springs Formation suggest that its maximum age is about 787 million years. This timing is just prior to the main rifting event in western Rodinia, suggesting that both the Crystal Spring and the Horse Thief Springs record, at most, only the initial subsidence and tectonism in the developing rift valley.

Above the Horse Thief Spring Formation lies the Beck Spring Dolomite. This formation is comprised of blue-gray to rusty-brown dolomite interbedded with varying amounts of fine sandstone, siltstone, and shale that typically weathers to an orange-brown color. In most localities near Death Valley, the Beck Spring Dolomite is about 1,000 feet thick and contains both microfossils of likely microbial organisms and laminations and other sedimentary structures resulting from the growth of colonies of such organisms. We will explore the nature of this microbial life in more detail in a later section, but the rocks and microfossils of the Beck Spring Dolomite are clearly of shallow marine origin. The interior basin in western Rodinia had once again subsided below sea level between 780 and 730 million years ago, and tidal inlets and lagoons developed in the localized submerged areas. However, this basin was restricted in extent and was not the product of a new, expanding ocean basin.

The main separation of Laurentia from the supercontinent, the event that formed the vast and enduring Great Basin seafloor, was yet to come. So was the main phase of the Snowball Earth climate disturbances. Both events are recorded in strata that overlie the Beck Spring Dolomite: a remarkable sequence of rocks within the Kingston Peak Formation and Noonday Dolomite.

The Ripping of Rodinia: Glaciomarine and Cap Carbonate Deposits

The Kingston Peak Formation was named by geologists in 1940 for the most prominent peak in the Kingston Range, southeast of Death Valley. There, the formation is about 1,200 feet thick and consists of a mixed assemblage of conglom-

FIGURE 2.5. A large, light-colored dolomite dropstone in fine-grained glaciomarine diamictite of the Kingston Peak Formation. Note that the laminations beneath the dropstone are deformed by the impact of the larger stone falling to the seafloor.

erate, sandstone, siltstone, dolomite, iron-rich siltstone, and what has been described as megabreccia (odd rocks comprised of large, angular chunks of different rock types). The formation also occurs across the Death Valley region in the Panamint Mountains (west of Death Valley), the Silurian Hills of the northern Mojave Desert, and in the Black Mountains east of Death Valley. Studies of these strata across this region have demonstrated that the Kingston Peak sediments are highly variable in thickness, sediment type, and stratigraphy from place to place. For example, in the Death Valley-Kingston Range area, the formation thickens southward from roughly 1,000 feet to nearly 9,000 feet.

Subdividing the Kingston Peak Formation in this region into consistent members has been challenging because the sediment comprising the formation is as variable as the thickness.

Each exposure of this rock unit looks a bit different in character from outcrops only a few miles away. This lateral variation in thickness and sediment type makes it difficult for geologists to precisely correlate strata within the formation from one exposure to another. Yet despite its variability and internal complexity, most geologists agree that Kingston Peak sediments provide the best record of the Neoproterozoic breakup of Rodinia and contemporaneous glaciations in western North America.

One of the few regional consistencies within the thick and varied pile of sediments in the Kingston Peak Formation is the presence of at least two major zones of very distinctive conglomerate known as diamictite. Diamictite is a poorly sorted sedimentary rock composed of a wide variety of particle sizes from large blocks of rock to microscopic grains (Figure 2.5).

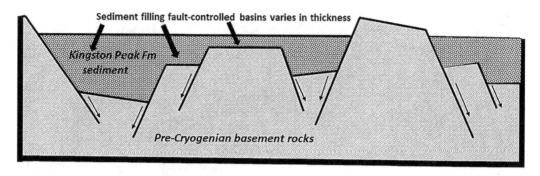

Sediment filling fault-controlled basins varies in thickness

Kingston Peak Fm sediment

Pre-Cryogenian basement rocks

FIGURE 2.6. Extensional faulting during deposition of the Kingston Peak Formation created local basins in which varying amounts of sediment collected.

Imagine boulder-sized chunks of rock accumulating along with pebbles, sand grains, fine silt, and mud in a single layer. After lithification, we would call the rock diamictite, like ordinary conglomerate (essentially, lithified gravel), but with an unusually large range of particle sizes. Across the Death Valley region, the diamictite zones in the Kingston Peak Formation vary in number, stratigraphic position, thickness, composition, and degree of metamorphism, but they are always present wherever reasonably complete exposures are examined. The diamictite occurs as layers or masses associated with other, more "ordinary," sedimentary rocks such as sandstone, dolomite, or siltstone.

Some of the diamictite and "megabreccia" deposits in the Kingston Peak Formation appear to have formed when landslides or gravity flows carried rocky rubble from elevated lands into subsiding basins flanking the highlands. These zones of coarse debris in the Kingston Peak Formation are commonly associated within marine sediments, indicating that the debris flows sometimes roared over the floors of bays and inlets as dense, turbulent, and fluidized masses of moving sediment known as turbidity currents. The rapid variations in thickness of the Kingstone Peak Formation appear to have been due, in part, to the active extensional faulting associated with rifting that produced the localized basins into which varying amounts of coarse sediments fell, perhaps following powerful earthquakes along the active faults (Figure 2.6). In fact, these ancient faults have been observed displacing Kingston Peak sediments in several places around the Death Valley region. The geometry and displacement of these faults identify them as the type produced by tensional stress (referred to as normal faults by geologists). Such stretching forces are consistent with the onset of continental rifting in this part of ancient Rodinia. From this evidence, it appears that the earliest stages of rifting involved extensional faulting that produced a complex pattern of highlands surrounded by flanking basins, some of which had fallen below sea level as Rodinia began to break apart.

There is strong evidence that some of the diamictites associated with marine sediments in the Kingston Peak Formation originated in a completely different manner. Larger stones embedded in some of the diamictite masses bear sets of small parallel grooves and scratches on their surfaces. These striations are very similar to those produced when rocks frozen into glaciers are ground against the bedrock beneath the mobile ice. In addition, the most extensive diamictite zones (those in what geologists call the Surprise Member of the Kingston Peak Formation) in the Panamint Range of Death Valley are some 800 feet thick and can be traced from outcrop to outcrop over more than 20 miles. Such widespread deposits are not likely the result of landslides or gravity flows, which are usually confined to channels, either on land or the seafloor. The diamictite in this region is interbedded with volcanic flows that have pillow structures, indicating undersea eruption

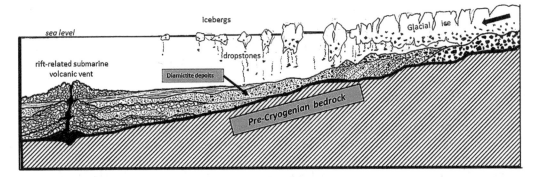

FIGURE 2.7. Origin of glaciomarine sediments. Unsorted sediment accumulates on the seafloor beneath melting glacial ice that moves from land to ocean in coastal areas.

and cooling of lava. However, very few of the rock particles comprising the diamictite units are of volcanic origin. Thus, the evidence suggests that these homogenous, widespread, and thick diamictite zones were deposited in a shallow marine setting by a mechanism that delivered great amounts of sediment of all particle sizes from distant sources and over broad areas.

Only one source seems likely to have dumped the rocky rubble containing the striated stones across the seafloor: the glaciers of Snowball Earth sliding over the basement rocks of Rodinia as it was breaking apart. The rift zones created during the disintegration of the supercontinent appear to have been evolving into new, volcanically active ocean basins into which coastal glaciers fell from land.

So compelling is the evidence of glacial origin for much of the Kingston Peak diamictite deposits that geologists have used the genetic term tillite for them. Till is a general term for all sediment dropped from melting glacial ice, while diamictite is simply a textural term describing sedimentary rocks comprised of a wide range of particle sizes. As we have seen, mechanisms such as gravity flows can also transport and deposit diamictite sediments but, when the rock particles fall from melting glacial ice over a broad area, then the term tillite is also appropriate. The Kingston Peak tillites are almost identical to modern sediments accumulating offshore in places like Norway or Alaska where glacial ice reaches the sea. Thus, the sediments comprising the Kingston Peak Forma-

tion are commonly described as glaciomarine in origin: they consist of glacially transported particles deposited on the seafloor from masses of melting sea ice (Figure 2.7).

Though the glaciomarine sediments in the Kingston Peak formation occur in several different horizons, there are two main portions where they are especially thick and prominent. The lower diamictite zone occurs mainly in the Surprise and Lime Kiln members and is thought to be slightly more than 700 million years old. This portion of the Kingston Peak Formation seems to correlate with the Sturtian glaciation of the Cryogenian Period. The middle portion of the Kingston Peak Formation is dominated by nonglaciomarine sandstone and shale, probably representing an interglacial epoch. Glaciomarine sediments reappear near the top of the Kingston Peak Formation (mostly in the Wildrose Member), and these deposits roughly correlate with the Marinoan glaciation, 650–630 million years ago. Moreover, similar glaciomarine and rift-related deposits can be traced from the Death Valley region into the Grand Canyon (the Chuar Group), Utah (Mineral Fork Tillite/Big Cottonwood Group and Perry Canyon/Dutch Peak Group), and Idaho (Pocatello Formation). These localities appear to define a more-or-less north–south zone of rifting that evolved into a similarly aligned seaway by the time of the Marinoan glaciation.

There is one more interesting twist on the story of continental break-up and climate oscillations recorded in the Cryogenian rocks of the Death Valley region. Immediately over-lying both zones of glaciomarine sediments in the Kingston Peak Formation are very distinctive sequences of dolomite or limestone, both chemical sedimentary rocks composed of the calcium and/or magnesium carbonate minerals precipitated from seawater. Except for the two occurrences immediately above the glaciomarine deposits, dolomite strata are relatively rare elsewhere in the formation. The lower diamictite is capped by the dark-colored Sourdough Limestone, while the upper glaciomarine sediments are overlain by the thicker (about 500 feet thick), creamy-tan Noonday Dolomite (Figure 2.9E). These two rock units are sometimes referred to as "cap carbonates" because they appear to form an abrupt cap on the underlying, and utterly dissimilar, glaciomarine strata.

The cap carbonates are interesting because precipitation of carbonate minerals occurs most readily in warm, clear seawater and is commonly facilitated by the algae or other photosynthetic organisms. During Snowball Earth conditions, when the water was cold and turbid, carbonate minerals would not form so readily in the marine environment. In addition, both cap carbonates show strong and abrupt negative shifts in the amounts of carbon-13 (^{13}C) incorporated into the rocks from the ancient seawater. This trend signifies significant and rapid warming of the Cryogenian seawater during the time the carbonate minerals were forming. The melting of glacial ice under warming conditions would have exposed rocks on land to weathering, including chemical reactions between silicate minerals in the bedrock and carbon dioxide, water, and oxygen in the atmosphere. Bicarbonate ions (HCO_3-) are released from these weathering reactions and would have stimulated the precipitation of $CaCO_3$ in the warming seas.

The two cap carbonates thus provide evidence for the abrupt end of the Sturtian and Marinoan glaciations, a relatively rapid warming of the seas, and the retreat of glacial ice. Some geologists have even speculated that the surface water during formation of the Cryogenian cap carbonate deposits may have been as warm as 40°C (104°F)! Because the contacts

between the glacial deposits and overlying cap carbonates tends to be fairly abrupt, the shift from frigid to sultry conditions appears to be geologically sudden.

These dramatic climatic oscillations during the time that Rodinia was breaking apart no doubt had significant effects on the environmental conditions across the developing oceanic basin, the nascent Great Basin seafloor. What makes these conditions especially fascinating is that from its earliest inception via the rending of a supercontinent—and beneath an ocean reeling from one extreme climatic perturbation to another—the Great Basin seafloor nurtured life. The creatures characterizing that early marine ecosystem, and how they might have been affected by the wild climatic upheavals at the end of the Cryogenian Period, are important considerations in understanding the evolution of marine ecosystems on early Earth. What do the rocks of the Great Basin tell us about this event? Let's have a look.

Life in the Cryogenian Period

No one knows for sure when or how life first emerged on our planet. However, most scenarios place the origin of living systems in the interval between about 4 billion years ago, when liquid water first condensed on the surface of the Earth, and about 3.5 billion years ago, the age of the oldest sedimentary structures likely to have been constructed by living organisms. In any case, the first organisms on our planet likely appeared in the primordial oceans billions of years before the initiation of the Great Basin seafloor. So, when the rift valleys of Rodinia first sank below sea level, the water that submerged them was not lifeless. It contained a rich biota of tiny unicellular creatures that had been evolving for billions of years. The rocks of Cryogenian age in and near the Great Basin provide some fascinating glimpses of these tiny microorganisms.

Fossils of tiny microorganisms have been identified from the Pahrump Group of the Death Valley region in carbonate and chert (microcrystalline silica) deposits of the preglacial Beck Spring Dolomite and in parts of the lower

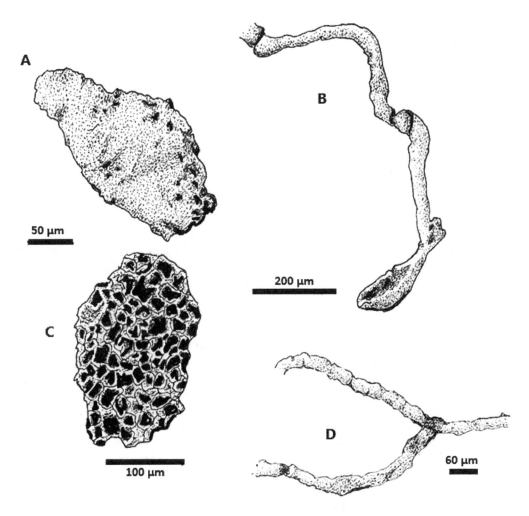

FIGURE 2.8. Cryogenian microfossils from the Pahrump group: (A) a vase-shaped cell similar to some modern amoebae; (B) long filamentous microfossil with an expanded terminus; (C) small microfossil with dimpled, possibly armored, surface similar to some modern testate amoebae; (D) a large branching filament similar to some modern cyanobacteria. All microfossils are from limestone or chert layers within the Kingston Peak Formation reported by Macdonald et al. (2013) and Corsetti and Kaufman (2003). Scale bars are in micrometers (μm), one millionth of a meter.

Kingston Peak Formation that also predate the earliest glacial deposits (Figure 2.8). Such fossils are invisible to the unaided eye and have been discovered through microscopic analysis of the masses and layers of chert. Recall from Chapter 1 that chert is a very fine-grained, nongranular sedimentary rock that occurs within the mostly carbonate rocks that contain them. These preglacial sediments probably accumulated in marginal marine basins adjacent to the rift valleys created as Rodinia began to break apart. However, there is also some evidence that the sediments became exposed due either to faulting and uplift of the restricted basins, diminishing sea level as Snowball Earth conditions commenced, or both from time to time during their initial deposition. The microfossils provide evidence that the Cryogenian seas, even though they lacked abundant oxygen and were probably stagnant and poorly circulating in the restricted basins, supported a surprisingly rich microbiota.

More than a dozen different types of micro-fossils are preserved in the Cryogenian strata of the Great Basin. Most are tiny filaments or spherical objects that appear to be the primitive (prokaryotic) cells of cyanobacteria, commonly known as blue-green algae (Figure 2.12). Such organisms in the modern world are photo-synthetic autotrophs ("self-feeding"), and their ancient ancestors probably were as well. Larger, generally vase-shaped cells have also been identified that are similar to modern testate ("shelled") amoeba, organisms that have big-ger and more complex (eukaryotic) cells. The larger unicellular animals were not autotrophic and thus provide evidence of at least a rudi-mentary food web in the Cryogenian seas. Rel-atively large and branching thread-like objects of likely organic origin have also been identified. These microfossils are generally interpreted as evidence of either cyanobacteria or colonies of more complex photosynthesizing algae. No truly multicellular organisms are represented in the Pahrump Group microbiota, but some of the fossilized cells appear to be clumped together into larger (but still microscopic) colonies.

Microfossils like those of the Pahrump Group also occur in coeval rocks of Utah and Arizona, suggesting that the microbial commu-nities were widely distributed in the Cryogenian seas that flooded the rift zones of Rodinia. More-over, the size and shape of the cells preserved as microfossils suggests that some of them may have been benthic (living on the seafloor) while others were planktonic (floating in the upper part of the water column). Such diversity and specialization of the various cell types suggests a complex microbial ecosystem and a lengthy evolutionary history prior to Cryogenian time.

Though no one knows precisely how the single-celled creatures lived or what modern groups of microbes they represent, if any, it is very likely that the majority were photosyn-thetic and planktonic, as are modern algae and cyanobacteria. This interpretation makes sense, because there is no fossil evidence that the marine biosphere of the Cryogenian Period was as complex and intricate as it is today. The many specialized ecological niches filled by larger multicellular creatures in the modern oceans simply didn't exist in the ancient micro-bial marine ecosystem. But regardless of the simplicity of the early marine food webs, the primary producers are still the essential founda-tion, and photosynthetic autotrophic organisms should have been the most abundant.

Another peculiarity of the Kingstone Peak sediments seems possibly related to the swarms of photosynthetic autotrophs in the Cryogenian seas. Mixed in with the conglomerate, diamictite, sandstone, and siltstone are relatively thin layers of laminated siltstone containing as much as 50 percent iron. The iron deposits are dominated by the mineral hematite, an iron oxide (Fe_2O_3) precipitated from seawater. At the time the iron deposits formed, the atmosphere contained less than half the oxygen it currently has. How did so much of the iron dissolved in seas become oxidized? Then and now, oxygen is a primary by-product of photosynthesis, and the primi-tive autotrophic plankton may have produced enough of it to help oxidize the iron washed into the sea from land or released from undersea lava flows. Once formed, the iron oxide minerals are insoluble and would accumulate on the seafloor along with other sediment.

An interesting twist on the fossilized microbes from the Pahrump Group is the obser-vation, first made by geologist Paul Knauth and his colleagues in the 1980s, that primitive micro-fossils, as well as geochemical signatures of life (carbon isotope ratios), also occur in pockets of cave rubble preserved within the Cryogenian carbonate rocks of the Pahrump Group. These deposits originated when the marine limestone and dolomite were exposed to surface condi-tions, and fresh water percolating through the exposed soluble rock dissolved cavities and pockets. This process is called karstification. Eventually, the caves and sinkholes formed through karstification were filled with rubble and silica-rich material washed in by surface streams and migrating groundwater. It was in the ancient cave sediments that Knauth discovered the fossil microbes and the geochemical anomalies that evince ancient organisms colonizing land areas. The cave deposits were depleted in carbon-13, which in the modern world occurs during photosynthesis by autotrophic organisms. Because the sediment was washed into the caves from the surface, the microfossils may have been living there as well. The fossils are very simple spheres or tubes, and scientists cannot be certain what organisms they represent. Nonetheless, they are similar in size and general appearance to cells and spores of modern cyanobacteria and other unicellular microbes.

This was, and still is, a fascinating discovery because it suggests that life may have originated on land far earlier than paleontologists previously assumed. The Neoproterozoic land surface has traditionally been envisioned by scientists as a barren and stark landscape where blistering radiation from the sun blasting through an oxygen-poor atmosphere all but sterilized the exposed rock. It wasn't until the Early Paleozoic Era, when atmospheric oxygen (and, hence, ozone) levels rose and the earliest plants became established on land, that life was thought possible without the protection of seawater. The Cryogenian microfossils from cave fillings in the *terrestrial* sediments of the Pahrump Group (and elsewhere, as was discovered later) are evidence that at least some organisms managed to adapt to the scorched surface of the Earth hundreds of millions of years before vascular plants appeared on land. Thus, as the Great Basin seafloor came into existence, the first oceans to cover it were teeming with microbial life. Even the shorelines may have glistened with a purple slime, as the first terrestrial life began to colonize land.

The Ediacaran Period: Microbial Menageries and the Advent of Complex Life

The final period of Neoproterozoic time, the Ediacaran, begins about 635 million years ago, after the last major glaciation of the Cryogenian. In the Great Basin and elsewhere, the advent of Ediacaran warmth is marked by the warm-water deposition of cap carbonates. The Noonday Dolomite of the Death Valley region is recognized as the cap carbonate sequence that signifies the end of the Cryogenian "Snowball Earth" conditions. Worldwide, there is evidence for a relatively brief glacial epoch, the Gaskiers glaciation, but this event is less extreme than the earlier cold intervals and is not so clearly recorded in the rocks of the Great Basin, unlike the older Cryogenian glacial episodes.

The Noonday Dolomite is one of the most distinctive rock formations of the Great Basin. In the Death Valley region, this formation is from 300 to 1,200 feet thick and generally forms bold cliffs and escarpments composed of light yellowish-gray to tan dolomite.

The dolomite is commonly laminated, bearing very thin wavy or wrinkled layers (Figure 2.9). In many exposures, the layers arch upward as much as several feet to form hemispherical masses known as stromatolites (Figure 2.9B). Both the wavy laminations and stromatolites in the Noonday Dolomite probably reflect the activity of microbes on the shallow seafloor. In fact, sedimentary rocks that exhibit such structures are known to geologists as "microbialites," sediments that accumulated through the combined activity of microbes and physical processes. Because we also will encounter microbialite fabrics elsewhere and at other times in the geologic history of the Great Basin, their origin is worth contemplating.

In many modern warm and shallow oceans, the seafloor is covered by a sticky organic mat composed of a bewilderingly complex microbial community that includes photosynthetic cyanobacteria and algae, archaea, fungi, and sometimes diatoms. The microorganisms in the mat are woven together as single cells, filaments, or clumps. Biologists commonly refer to these mats as "biofilms" because the microbes are contained within a sticky, organic, slime-like substance. Despite the popular conception of such organic slimes as primitive, the community structure of the biofilms is complex and intricate, with each microbe interacting with the others and with the environment to sustain a robust microecosystem. Sediment that accumulates on the seafloor is trapped by the biofilm like insects on flypaper. Simultaneously, chemical reactions

FIGURE 2.9. Features of microbialite deposits: (A) wavy laminations resulting from successive growth of microbial mats; (B) dome-shaped stromatolite with internal lamination; (C) thrombolite clotting produced by dark-colored, unlaminated, branching structures; (D) oncolites, small oval-shaped nodules with concentric lamination; (E) close-up the wavy lamination of the Noonday Dolomite in the Panamint Range, Death Valley Region. Scale bar in all photos = 5 cm (2 in.)

between the ions dissolved in seawater and the organic compounds in the biofilm can lead to precipitation of carbonate minerals on the slimy mat. This is especially true of the photosynthetic cells (such as cyanobacteria) living in the mat. Photosynthesis consumes carbon dioxide from the seawater, which in turn reduces the acidity (technically, raises the pH), prompting the precipitation of calcium carbonate minerals in and over the mat. Eventually, sediment covering the biofilm blocks sunlight from reaching the photosynthetic microbes, and the microbial ecosystem collapses. However, reproductive spores released to the water settle on the dormant mat surface, and the colonies are regenerated to trap additional sediment. Many repetitions of this cycle lead to the accumulation of finely laminated sediment (Figure 2.9A, E).

Over time, the original organic matter in the sediment is replaced with calcite, aragonite, dolomite, or even silica. The resulting rock is typically a hard limestone or dolomite bearing the original laminations produced by microbial activity on the seafloor—a microbialite. Aside from their characteristic lamination, microbialite deposits commonly preserve other structures resulting from the growth of photosynthetic microbes. For example, stromatolites are internally laminated, hemispherical structures built

upward from the seafloor, sometimes forming columns as much as several feet tall (Figure 2.9B). Thrombolites are knobby, sometimes branching structures constructed by photosynthetic microbes but that typically lack the lamination characteristic of stromatolites. Carbonate sediment accumulating between adjacent thrombolites can result in a limestone or dolomite with a clotted appearance (Figure 2.9C). When shallow seawater is agitated by waves, bottom currents, or tidal motions, the microbial activity commonly results in oval or nearly spherical bodies with internal lamination known as oncolites (Figure 2.9D). Beneath shallow, warm seas, where conditions for the growth of cyanobacteria and other microbes are optimal, these structures can grow in such profusion that masses like modern reefs can develop. These reef-like masses have been identified in the Ediacaran microbialite deposits of the Deep Spring Formation of western Nevada.

The Late Ediacaran: Wormworld and a Passive Margin

In latest Ediacaran time, after the deposition of the Noonday Dolomite in the Death Valley region, the rifted edge of Laurentia became fully separated from other parts of Rodinia in the Great Basin region. This new continental

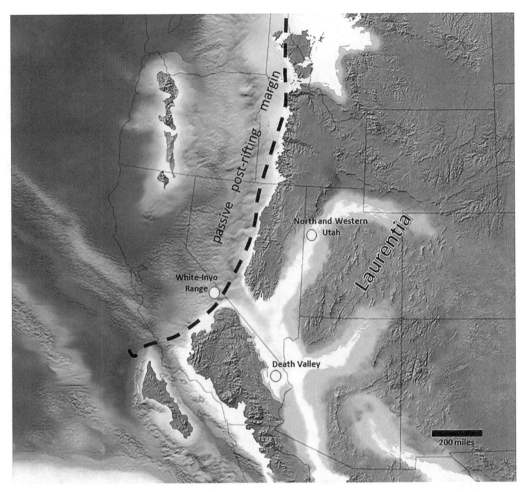

FIGURE 2.10. Paleogeography of the Great Basin Region during the Ediacaran Period, 636–600 million years ago. Paleogeographic base map prepared by R. Blakey, Colorado Plateau Geosystems Inc.

margin, formed from the rending of a supercontinent, was bordered by a narrow seaway aligned roughly north–south. This seaway developed from one of the submerged and integrated rift zones that originated earlier in the Neoproterozoic Era. On the floor of the nascent ocean basin was an active oceanic ridge where seafloor spreading was creating new lithosphere for the plate carrying Laurentia to the east (relative to modern geographic orientations). The rifted margin of Laurentia bordering the narrow ocean basin extended for thousands of miles in a roughly north–south direction. Seafloor spreading continued along the newly initiated ridge system to the west, and the rocks of the

Laurentian plate cooled and contracted, becoming denser as they drifted farther east from the axis of the undersea volcanic ridge. This led to the sinking, or subsidence, of the lithosphere beneath the earliest Great Basin seafloor and a great flood of sand, silt, and mud was washed from land into the narrow, deepening seaway. These materials covered the older cap carbonates, such as the Noonday Dolomite, under great westward-thickening wedges of granular sediments (Figures 2.10, 2.11).

In the Death Valley region, the Johnnie Formation (a mix of siltstone, fine sandstone, sandy limestone, and thin dolomite layers as much as 10,000 feet thick) covers the Noonday

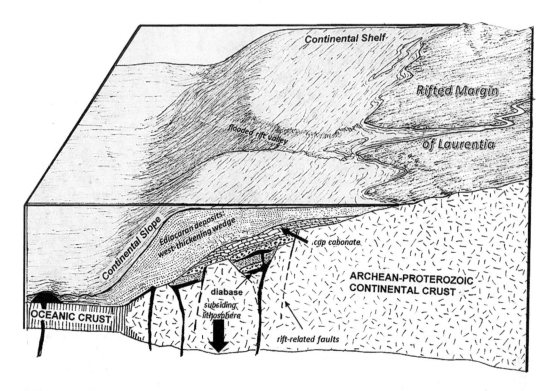

FIGURE 2.11. The passive margin of Laurentia during Ediacaran time. Sediments accumulated on the rifted edge of the continent in west-thickening sequences in near-shore to offshore settings. Active subsidence of the continental margin allowed thousands of feet of sediment to accumulate under water that was only hundreds of feet deep.

Dolomite in most places. The sediments of the Johnnie Formation were deposited in a variety of sedimentary environments along the margin of Laurentia, including lagoons, embayments, tidal flats, sandy beaches and offshore sandbars, and deltaic streams. In the upper portion of the Johnnie Formation, there is evidence of a period of exposure and erosion that produced incised valleys and channels partially filled with stream-deposited conglomerate and sandstone. This brief period of exposure may coincide with a sea-level drop that accompanied the Gaskiers glaciation, a relatively brief event in the Late Ediacaran Period. Above the Johnnie Formation is the Sterling Quartzite, a formation dominated by medium-to-coarse sand that was deposited in shallow offshore and nearshore environments. Overlying the Sterling Quartzite is the lower part of the Wood Canyon Formation, consisting of siltstone, shale,

and sandstone. These three formations span a time interval in the Late Ediacaran Period from about 630–550 million years ago. They provide an almost continuous record of events and conditions immediately after the rifting of Rodinia was complete.

In addition to the Ediacaran strata of the Death Valley region, rocks of this age are widely distributed in other Great Basin regions. In western and central Utah, similar deposits occur in the Inkom, Mutual, and Browns Hole Formations, though land-deposited sediments appear to be more dominant than marine sediments in these units. In the White-Inyo Mountains along the Nevada-California border north of Death Valley, Ediacaran sediments comprise the bulk of the mostly marine Wyman, Reed, and Deep Spring Formations. Ediacaran strata also extend north from the Great Basin into Idaho, where they are recognized as parts of the Brigham

Group. The broad distribution indicates that the seaway bordering the rifted margin of Laurentia extended thousands of miles generally north–south, with several major inlets and bays where rivers discharged sediment (Figure 2.11). Geologists commonly refer to this general tectonic setting as a passive margin, meaning that both the continental and adjacent oceanic lithosphere reside on a single plate. The modern east coast of North America is a good example of a passive margin. There, the continental crust of North America and the oceanic crust of the Atlantic Ocean are both carried westward on the North American plate by seafloor spreading at the mid-Atlantic ridge system. Passive margins are geologically quiet settings where few strong earthquakes or violent volcanic eruptions occur. In the Late Ediacaran, sediment washed from land and generated within the seas accumulated on the passive margin of Laurentia in westward-thickening wedges (Figure 2.11), with near-shore or land-deposited material grading into deeper-water deposits. This general pattern of passive margin sedimentation was well established in Ediacaran time and persisted for some 350 million years, well into the Paleozoic Era.

The Dawn of Wormworld

As we have seen, the microfossils and microbialite deposits of the Cryogenian Period document a diverse and complex microbial biota in the earliest seas that submerged the Great Basin seafloor. Both simple (prokaryotic) and more complex (eukaryotic) cell types were present in these seas as plankton, in seafloor microbial mats, and perhaps as slimy masses along the shore and on land. Though these microbial colonies could sometimes construct large structures such as stromatolites and thrombolites, there are only a few hints, mostly loose aggregations of individual cells, in the Cryogenian fossil record of truly multicellular organisms. It seems that multicellular animals (metazoans) and plants (metaphytes) were either absent from the Cryogenian seas or were too rare to leave abundant fossils.

All of that changed in dramatic fashion in the Early Ediacaran Period. Based on fossil evidence

worldwide, it appears that about 600 million years ago, the seafloor exploded into complex multicellular life, supporting not just a few tiny creatures but a rich and diverse fauna of large metazoans and metaphytes unlike anything that preceded or followed it. The Ediacaran organisms were bizarre and strange creatures, so unlike modern sea life that many of them cannot be easily related to living descendants. Collectively, they comprise a marine fauna so unique that paleontologists refer to it as the Ediacaran biota. The period of geologic time distinguished by these odd fossils was named in 2004 after the Ediacara Hills of Australia, where the remains are particularly abundant. Discoveries of additional fossils in recent years have added much to our understanding of the Ediacaran biota, but many questions remain concerning the organisms, the ecosystems that supported them, and their patterns of evolution. Without a doubt, the Ediacaran Period is one of the most intriguing episodes in the long history of life on our planet.

Ediacaran fossils are typically preserved in medium- to fine-grained sediments such as siltstone and shale, usually as impressions in the fine sediment. None of the larger Ediacaran organisms appear to have possessed mineralized skeletal material. The earliest fossils of this assemblage are delicate, frond-like forms that are preserved in relatively deep-water sediments. In younger Ediacaran strata, the fossils are found in both granular and carbonate sediments, commonly with microbialite features. So, it is plausible that the Ediacaran creatures may have originated in deeper water below the photic zone but soon expanded their range into the shallow seas, where photosynthetic microbes were abundant. In many localities, Ediacaran fossils are preserved along with the tracks and trails (trace fossils) left by unknown metazoans plowing through the mud and microbial mats of the shallow seafloor. The trace fossils were probably made by flexible, worm-like organisms churning through the sediment or crawling along the muddy bottom. The soft nature of the creatures represented by most Ediacaran body and trace fossils has led some geologists to refer to this biota as "wormworld." Though it is well beyond the scope of this book to detail all the creatures comprising the global Ediacaran wormworld biota, we can produce a fair picture of this extraordinary assemblage by describing a few of them.

The most complete and best-studied record of Ediacaran events and life in the Great Basin lies in strata exposed in the southwestern part of the province, near the California-Nevada state line. Here, the Ediacaran sediments comprise most of the Johnnie and Sterling Formations of the Death Valley region and the lower portion of the overlying Wood Canyon Formation. In the White-Inyo Mountains to the north, Ediacaran strata are present in the upper Wyman Formation, Reed Dolomite, and lower portion of the overlying Deep Spring Formation. Ediacaran sediments also occur in western Utah and southern Idaho, but fewer fossils have been recovered from these strata.

Like most other fossils of this age, the Great Basin forms are preserved as impressions in siltstone and shale, with little evidence of hard, mineralized skeletons (Figure 2.12). *Conotubus* is a small, slender conical structure, an inch long or less, with circular ribs. *Gaojiashania* is similar, but larger, with many closely spaced ribs giving the appearance of a segmented tube. *Nimbia* fossils are ring-shaped structures, about 2 inches in diameter, that appear to have possessed thickening rims somewhat like modern jellyfish. *Ernietta* is evidently shaped like a haystack 1 to 3 inches wide, with soft, tentacle-like segments. In life, this creature may have looked a little like a small sea anemone. While the larger Ediacaran organisms appear to be completely soft-bodied, some smaller fossils that accompany them do seem to have possessed mineralized skeletons. *Cloudina,* the most common of these fossils in the Great Basin, had a tiny (less than 1 inch long), calcareous shell comprised of stacked funnel-shaped cones.

Because the Ediacaran creatures are so unlike anything in the modern seas, scientists are uncertain about any relationship with living groups of marine invertebrates. The evident segmentation of many Ediacaran fossils, along with their burrowing behavior, seem to suggest a relationship with one or more of the several phyla (such as Annelida, Chaetognatha, or Hemichordata) of living "sea worms." Some Ediacaran forms may be distantly related to sponges in the phylum Porifera. Others could be ancestral to corals and jellyfish of the phylum Cnidaria. Radial symmetry appears in some Ediacaran forms, suggesting a relationship to sea urchins and starfish of the phylum Echinodermata. Some scientists have even proposed that certain Ediacaran fossils may be lichens, fungi, or seaweeds. If so, the Ediacaran biota includes forms related to modern photosynthetic marine plants and phytoplankton.

Aside from their uncertain relationship with modern groups of sea life, several other fundamental questions also remain about the Ediacaran biota. How, and what, did they eat? How did they reproduce? Which were sessile and which were mobile? Could they swim, or were they restricted to squirming and crawling? How did they sense their environment? What factors led to the evolution of large, multicellular organisms from microbial ancestors? We are still seeking answers to these questions, and controversy about the nature of the Ediacaran biota abounds among paleontologists. Nonetheless, the fossils tell us one thing for sure: starting around 600 million years ago, specialized metazoans and/or metaphytes exploded across the Great Basin seafloor and elsewhere in the global ocean. These multicellular creatures probably arose from unicellular ancestors much earlier in Neoproterozoic time and became adapted to a range of ecological niches through a long evolutionary process.

By the end of the Ediacaran Period, many lineages of highly specialized, soft-bodied creatures had evolved distinctive body plans and behaviors well suited to the muddy bottom of the seafloor. Many appear to have been suspension feeders, consuming plankton that descended to the seafloor. The conical fossils could have been suspension feeders, their mineralized exoskeletons allowing them to rise upward from the murky bottom to intercept more food and nutrients. The abundant trace fossils evince worm-like deposit-feeding animals that processed

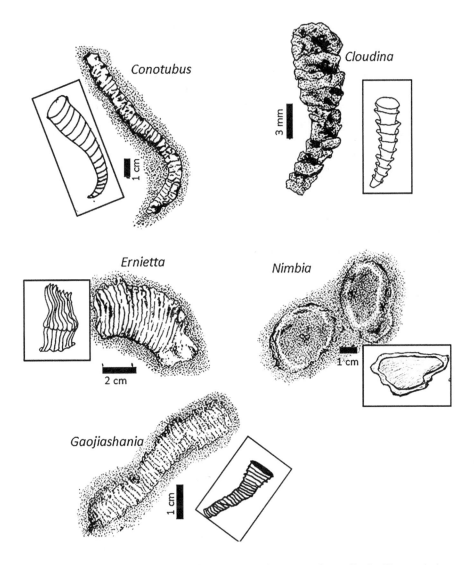

FIGURE 2.12. Sketches of Ediacaran fossils from the southwestern Great Basin. The scale bars express their small size, and the figures in the boxes depict the overall morphology of the organisms.

organic matter from the cyanobacterial mats and surrounding mud. Scientists have also discovered *Cloudina* fossils with small holes drilled into them by an unknown predator. (The mineralized exoskeleton of *Cloudina* may even have been an evolutionary response to such predation). All of this suggests a rather complex marine ecosystem with multiple ecological niches filled by a diverse array of organisms, a major shift in Neoproterozoic sealife. For the

first time since the origin of life billions of years earlier, the ocean floor became blanketed with large organisms, diverse in form and ecological function. Competition for survival among the various groups stimulated rapid evolution that permanently reshaped the life of the seas. Once this ecological threshold was passed, there was no turning back. The Great Basin seafloor would never be the same after the Ediacaran revolution.

The Ediacaran biota persisted for about the final 30 million years of the period, during which the individual organisms and the structure of the seafloor communities they comprised experienced steady evolution. As the Ediacaran Period drew to a close 545 million years ago, at least two periods of global extinction swept across the shallow seafloor, followed by the proliferation of surviving and newly evolved forms alike. This was also a time of significant biogeochemical instability in the global ocean, with erratic swings of carbon isotope values, alternating oxygenation and stagnation of deep seawater, and changes in water chemistry. Such shifting conditions were strong influences on the ever-changing Ediacaran biota. Paleontologists are still working out the detailed patterns of extinction and evolution among the enigmatic Ediacaran organisms, but there is near-universal agreement that the final extinction, about 540 million years ago, wiped out nearly all these peculiar metazoans.

This final extinction set the stage for an even more dramatic change on the Great Basin seafloor: the detonation of the "Cambrian explosion" that marks the inception of the Phanerozoic Era. After that event, life on the ocean floor was once again utterly restructured—and would stay that way for the next 50 million years.

3

The Great Cambrian Explosion

The Cambrian Period begins about 541 million years ago as the initial period of the Paleozoic Era, the earliest of the three subdivisions of the Phanerozoic Eon. The Phanerozoic Eon, literally the eon of "visible life," is characterized by abundant fossils of large marine organisms, most of which can be associated one way or another with modern relatives. All the major groups (phyla) of marine invertebrates became well established, if not abundant, in the Cambrian Period, and the evolutionary history of most of them can be traced through the succeeding periods of the Phanerozoic Eon. However, as we will see in our exploration of Cambrian life on the Great Basin seafloor, several aspects of this prehistoric fauna make it unique compared to younger Paleozoic assemblages. First, let's set the stage for our Cambrian adventure by revisiting the passive margin of Laurentia.

We have seen that the western edge of Laurentia developed during Neoproterozoic rifting, and thick sequences of sediment began to accumulate on the passive margin in west-thickening wedges. This tectonic setting persisted into Cambrian time with no major modification. Shallow marine sediments continued to accumulate over the underlying Ediacaran deposits in the same general pattern established after the rifting of Rodinia. The Great Basin region on the western side of modern North America, was situated along the edge of Laurentia that then faced north and was located within a few degrees of the equator 540 million years ago.

In this low-latitude location, the climate was likely warm and tropical, and ocean currents like those along the equator today would have swept east to west (north to south, in modern orientation) across the shallow seafloor. At the beginning of Cambrian time, the shoreline of Laurentia was in western Utah and southern Nevada, and from there the seafloor descended gently westward hundreds of miles toward the edge of the continental shelf. The water over the shallow seafloor was clear, warm, and well-agitated by currents and waves.

Beyond the shelf edge, in east-central California, the Great Basin seafloor dropped abruptly into deeper water where dark ooze and mud accumulated in the murky depths. Nearly all of what is now the Great Basin in Cambrian time could be envisioned as a serene, tropical paradise similar in general character to the warm seas of modern Polynesia or the Caribbean—except, of course, without the swaying palm trees or any other terrestrial plants for that matter. Inland from the Cambrian coast, the surface of Laurentia was a nearly sterile rocky plain. The multiple glaciations of the Cryogenian (and probably one additional, less extensive one in the Ediacaran) had bulldozed great expanses of Laurentian basement rock into a low, rolling plain with, at most, a few undulating hills.

Ever since the separation of Laurentia from Rodinia, the passive margins surrounding it had been subsiding under the weight of sediments

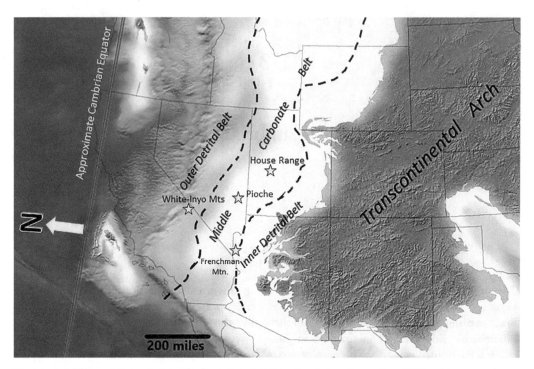

FIGURE 3.1. The passive margin of Laurentia in Middle Cambrian time, about 510 million years ago. Paleogeographic base map prepared by R. Blakey, Colorado Plateau Geosystems Inc.

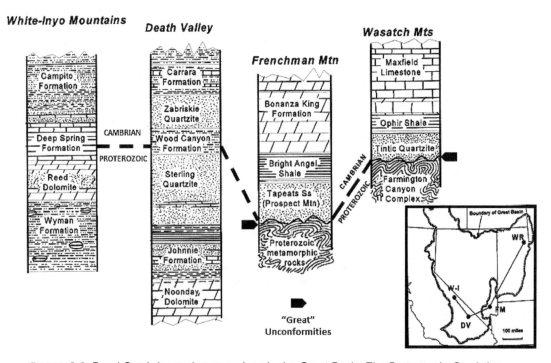

FIGURE 3.2. Basal Cambrian rock successions in the Great Basin. The Proterozoic-Cambrian boundary is conformable within thick rock successions in the western part of the region, while a distinct unconformity (technically, a disconformity) separates Cambrian deposits from much older basement rocks in the east.

FIGURE 3.3. The Great Unconformity of the eastern Great Basin: (A) close view of Cambrian sandstone resting on Proterozoic basement rocks at Frenchman Mountain in southern Nevada; (B) tilted layers of light-colored Cambrian sandstone lying atop dark-colored metamorphic basement rocks in the northern Wasatch Range, near Brigham City, Utah. Dashed yellow line marks the Great Unconformity in both photographs.

eroded from the interior and deposited over the rifted crust. As the newly formed oceanic lithosphere cooled and contracted, it sank even faster due to its increased density. By Cambrian time, the subsiding margins surrounding Laurentia had depressed its edges so much that only an elongated oval region in the center of the continent remained above sea level, unburied by sediment. This exposed region of post-rifting Laurentia has been named the Transcontinental Arch (Figure 3.1). After the multiple glaciations of the Cryogenian Period, meltwater that returned to the sea from terrestrial ice caused a global rise in sea level beginning in Ediacaran time. In the Cambrian Period, this sea level rise was enhanced by an of acceleration seafloor spreading in newly formed ocean basins. Some geologists have estimated that sea level rose by as much as one foot every 20 years in Early Cambrian time. Without high mountains to block the stampeding seas, the Laurentian shorelines migrated thousands of miles inland

over the coastal lowlands. By the end of Cambrian time, the rising sea had inundated about 75 percent of Laurentia and only the higher parts of the Transcontinental Arch remained exposed. Though this great transgression began in Late Ediacaran time, the Cambrian Period sedimentary record throughout North America reflects the rise in sea level most dramatically. Sediments that accumulated during this drowning of Laurentia have been named the Sauk sequence, the earliest of several major transgressive rock sequences identified in the Paleozoic strata of North America.

The Sauk transgression is well documented in the Cambrian strata of the Great Basin. In the western part of the region, the Great Basin seafloor was already deeply submerged at the beginning of Cambrian time, and the rise in sea level had little effect on the type of sediment that accumulated on the seafloor. In this area, the Ediacaran-Cambrian boundary is difficult to identify based on rock type alone

(Figure 3.2). Fossil evidence suggests that the Ediacaran-Cambrian boundary here is located within the sandstone and mudstone layers of the Wood Canyon (Death Valley area) and Deep Spring Formations (White-Inyo Range).

However, in the eastern Great Basin, as the shoreline migrated east during the Sauk transgression, the advancing seas submerged deeply eroded metamorphic and igneous bedrock that had been exposed as dry land ever since the Cryogenian glaciers receded. Sand and gravel were deposited in coastal and shallow offshore settings above Precambrian crystalline rocks hundreds of millions of years older. In places such as Frenchman Mountain or the Wasatch Range, the dissimilarity between the Cambrian sandstone and the much older basement rock below makes the Cambrian-Precambrian boundary glaringly obvious as a type of unconformity known as a nonconformity. So dramatic is this boundary that it has earned the informal appellation "Great Unconformity" in many places where it is well exposed (Figure 3.3). Rocks below the Great Unconformity range in age from about 1.7 billion years to perhaps more than 2 billion years. Thus, the Great Unconformity represents a gap in the rock record of more than a billion years. Because sediment deposition was nearly continuous from Late Cryogenian to Cambrian time in the western Great Basin, the Great Unconformity dies out in that direction.

Throughout the Cambrian Period, the Sauk transgression resulted in the eastward march of the shoreline deeper into the interior of Laurentia. This encroachment resulted in the complete inundation of the entire Great Basin region by Middle Cambrian time. So persistent was the Sauk transgression, that the seas would not begin to recede from central Laurentia until early in the Ordovician Period (about 470 Ma). During the long period of complete submergence, thick sequences of sediment accumulated on the Great Basin seafloor. In the House Range of western Utah, for example, more than 12,000 feet (over 2 miles!) of sedimentary rock accumulated, mostly in the middle and late parts of the Cambrian. To the west, in the White-Inyo Mountains of eastern California, the Cambrian

sequence is even more impressive, roughly twice as thick as in the House Range, and consists of sediment deposited on the passive margin throughout the entire period.

Following the inundation of the Great Basin seafloor after the Sauk transgression, sediment accumulated on the passive margin of Laurentia in a consistent pattern for the remainder of Cambrian time. Near the shoreline, silica-rich sand, silt, and mud, derived mostly from the erosion of the Transcontinental Arch, was deposited in relatively shallow water. These sediments comprise the sandstone, shale, and mudstone of the inner detrital belt (Figures 3.1, 3.4). Farther offshore, but still in water less than about 500 feet deep, thick sequences of carbonate mud and silt accumulated. The carbonate rocks that formed from this sediment constitute the middle carbonate belt, dominated by thick sequences of massive limestone and dolomite with microbialite features such as stromatolites and fossils of Cambrian marine organisms. The outer detrital belt is generally seaward from the middle carbonate belt; there, thin-bedded, dark-colored limy silt, mud, and ooze accumulated in deeper-water settings. The rocks of the outer detrital belt sometimes contain abundant fossils but typically have fewer microbial features than the those of the middle carbonate belt. This may reflect the limiting effect of diminished light on photosynthetic cyanobacteria and algae living on the deeper seafloor. In the central and western Great Basin, rocks of the outer detrital have commonly been deformed by post-Cambrian geological events (Figure 3.4)

Cambrian Rock Sequences in the Great Basin

The Cambrian strata exposed in the mountains of the Great Basin accumulated over a span of more than 50 million years. Granular rock types such as sandstone and siltstone are dominant in the relatively thin Cambrian sequences of the inner detrital belt of the eastern Great Basin. Rock units such as the Tintic Quartzite, the Prospect Mountain Quartzite, and the Tapeats Sandstone are typical of the coarse, sandy sediments that were deposited near the eastward-migrating shore line. These sandy units are

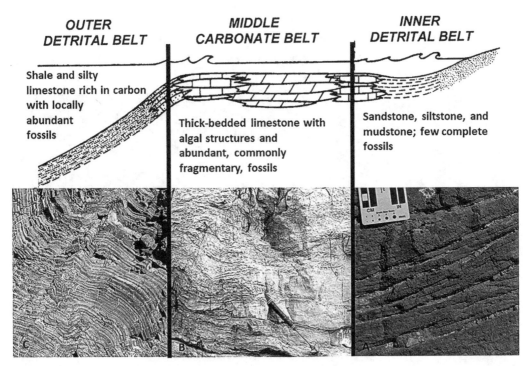

OUTER DETRITAL BELT

MIDDLE CARBONATE BELT

INNER DETRITAL BELT

Shale and silty limestone rich in carbon with locally abundant fossils

Thick-bedded limestone with algal structures and abundant, commonly fragmentary, fossils

Sandstone, siltstone, and mudstone; few complete fossils

FIGURE 3.4. The post-Sauk transgression pattern of Cambrian rocks in the Great Basin region. Outcrop views of rock types typifying the three belts are (A) Tapeats Sandstone at Frenchman Mountain, Nevada; (B) Lynch Dolomite in the Lakeside Mountains, Utah; (C) deformed strata of the Swarbrick Formation, Hot Creek Range, Nevada.

associated with and overlain by shaly intervals such as the Ophir Shale and Pioche Formation in much of western Utah and eastern Nevada. It is not uncommon to observe current crossbedding and ripple marks in these granular rocks, suggesting that swift and persistent bottom currents swept the shallow seafloor.

Westward from the inner detrital belt, limestone and dolomite become more common and are the dominant lithologies of the thick successions of Cambrian strata of western Utah and eastern Nevada. Dozens of different formations have been established within the extremely thick limestone-dominated middle carbonate belt (Figure 3.5). Many of these carbonate rocks contain well-preserved stromatolites, thrombolites, microbial laminations, and abundant fossils that indicate deposition in clear, well-illuminated water probably less than about 500 feet deep (Figure 3.6). It may seem odd that more than 12,000 feet of carbonate sediments could collect in only a few hundred feet of water.

Such thick accumulations of shallow-water sediments are possible because the passive margin of Laurentia was steadily subsiding throughout Cambrian time. For every foot of carbonate mud that settled onto ocean bottom, the floor of the Great Basin seas sank to accommodate it.

In the far western part of the Great Basin, the very thick Cambrian succession is a mixed assemblage of deeper-water shale, siltstone, and sandstone, with intermittent limestone sequences. Some exposures of Cambrian rocks in the western Great Basin have abundant burrows and tracks (trace fossils) excavated by soft, worm-like organisms churning over and through the soft ooze (Figure 3.7). In fact, one of these trace fossils known as *Treptichnus pedum* evinces very sophisticated burrowing behavior and is used worldwide to formally identify the beginning of Cambrian time in the rock record. Because deposition of sediment was continuous in the western Great Basin from Late Ediacaran through Cambrian time, the strata exposed in

FIGURE 3.5. Cambrian strata exposed on the western face of the House Range, Utah. The prominent peak is Notch Peak (elevation 9,654 feet), a landmark visible for miles in the West Desert of Utah. The Cambrian succession in the House Range is more than 12,000 feet thick and is dominated by limestone and dolomite in several different formations.

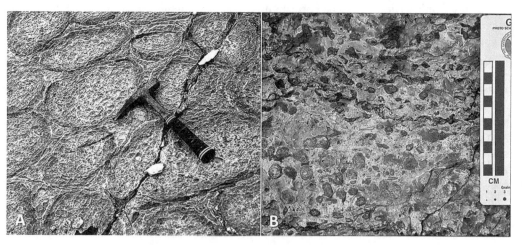

FIGURE 3.6. Microbial features in Cambrian rocks of the Great Basin: (A) bedding plane view of stromatolites in the Lynch Dolomite, Lakeside Mountains, Utah. Circular objects are cross-sections of the hemispherical structures built by photosynthetic microbes. (B) Oncolites, or nodular stromatolites, from the Bonanza King Formation, Nopah Range, California. Small, rounded objects represent microbially trapped sediment masses that form in water agitated by oscillating currents. Smallest scale bar divisions = 1 cm.

FIGURE 3.7. Trace fossils from the Cambrian Campito Formation, White-Inyo Range, eastern California. These marks were made by soft-bodied organisms squirming through the silty ooze and processing the sediment for organic matter. Inset photograph show the coarse, burrow-filling sediment in cross-sectional view. Smallest scale bar divisions = 1 cm.

the desert mountains of eastern California are world famous as an unbroken record of the events that accompanied the transition from the Proterozoic Eon to the Phanerozoic Eon. Yet, while the subsiding passive margin of Laurentia remained geologically serene through this phase of geologic time, the same cannot be said of life on the seafloor.

The Ediacaran Extinction and Cambrian Explosion

The advent of the Cambrian Period and the Phanerozoic Era was marked by dramatic changes in the communities of creatures populating the Great Basin seafloor. In the final few million years of Ediacaran time, the strange, soft-bodied fauna of "wormworld" experienced a severe, pulsed extinction. The first wave of extinction swept across the deeper seafloor

about 550 million years ago. This event was followed by the abrupt disappearance of most of the remaining Ediacaran animals a few million years later, just as the Cambrian Period dawned. The cause of the Ediacaran extinctions is a topic of considerable controversy among paleontologists. There is geochemical evidence (uranium isotopic composition) for a marked decline in oceanic oxygen levels about the same time as the first extinctions began. However, there is also a general increase in the abundance and diversity of trace fossils and a roughly simultaneous decrease in the dominance of microbialite deposits as the extinctions transpired. This suggests to some scientists that the sediment-grazing and churning organisms may have contributed to the decline of the Ediacaran biota by destabilizing the microbial mat substrates to which they were well adapted. Recall, also, that

FIGURE 3.8. The typical form of *Treptichnus pedum* trace fossils from the earliest Cambrian deposits. The arrows indicate the inferred systematic movement of a probing, deposit-feeding organism.

some of the small, shelly Ediacaran fossils, such as *Cloudina*, show signs of predation by some unknown large marauder. Perhaps many other soft wormworld organisms also suffered from predation but left no fossils to document the cause of their demise.

Like most great extinctions in earth history, the collapse of the Ediacaran biota may reflect multiple sources of stress, both biological and environmental, that in combination doomed so many organisms. While the cause of the Ediacaran extinctions remains debatable, evidence from the fossil record clearly reveals a major shift in the character of life on the ancient seafloor. This biotic transformation set the stage for what has been called the "Cambrian explosion," during which time marine faunas became strikingly more abundant, diverse, and utterly dissimilar to their Ediacaran predecessors.

The resulting explosion of fossils is a real phenomenon, but we should note that it does not occur at the inception of Cambrian time. By international agreement, the base of the Cambrian rock record is defined by the first occurrence of a unique trace fossil known as *Treptichnus pedum* (Figure 3.8). These structures were excavated by an unknown, worm-like, deposit-feeding organism that plowed through the soft sediment in a complex branching or "feather-stitch" path. The pattern of branching in this trace fossil records the systematic probings of a soft-bodied organism literally mining the sediment for bits of organic matter. The relatively sophisticated movement implied by *Treptichnus* and other Early Cambrian trace fossils suggests that trace-making organisms were more complex, capable of greater mobility, and possessed of more sensitive sensory abilities than the "worms" that preceded them. We find a much more robust and diverse set of invertebrate creatures churning through the mud of the Great Basin seafloor after the extinction of the Ediacaran fauna.

The explosion of deposit feeders roiling the Early Cambrian ooze converted the seafloor from crusty microbial mats to soft mud constantly reworked by intense burrowing activity. Thus, the initial burst of life in the Cambrian Period was in the mud and muck with a rapidly evolving new fauna of deposit-feeding organisms. Unfortunately, few body fossils other than tiny, shelly creatures, holdovers from the Ediacaran fauna, are found in the earliest Cambrian rocks of the Great Basin.

Ten to 20 million years after the Cambrian Period began, however, there was a second explosion of life, this one much more obvious in the worldwide fossil record. Almost immediately, mineralized skeletal material from all the major lineages (phyla) of marine invertebrates appears. It seems that hard and durable tissues covering or imbedded in the soft flesh of marine organisms evolved independently and simultaneously in many different groups of previously unarmored marine invertebrates. Though the Cambrian explosion of the fossil record has been well known to paleontologists for more than two centuries, there is still no perfect explanation for why hard parts evolved almost concurrently in so many unrelated lineages.

It may be that ocean chemistry, particularly the concentration of phosphorus, calcium, and magnesium, changed in ways that made it possible for organisms to secrete calcareous and phosphatic compounds from seawater. The explosion of the fossil record does occur during the time that the Sauk transgression was submerging the mineral-rich soil and bedrock of previously exposed land surfaces. Ions generated during the weathering of these rocks might have been released into the seas, raising the concentration of dissolved elements suitable for building skeletal material. Perhaps it was the rapid evolution of grazing and predatory organisms that propelled an evolutionary arms race among both the pillagers and their prey. The oxygenation and mixing of bottom muds, along with the destruction of the bacterial mats by burrowing organisms may have created new ecological niches on the Cambrian seafloor that favored the development of hard parts. Whatever the causes, this drastic increase is also well preserved in the Cambrian rocks of the Great Basin.

However, as we will soon see, unique conditions of preservation in several places on the Great Basin seafloor provide evidence that this dramatic pulse of evolution affected even the groups of soft-bodied organisms that did not possess shells. Thus, the Cambrian explosion in the fossil record is not just the result of the evolution of more easily preserved hard skeletal material. The notorious Cambrian explosion can, in fact, be thought of as three approximately coeval explosions that occurred early in the period. There is good evidence to support the view that within the first 10 or 20 million years of the Cambrian Period there were sudden increases in the diversity and abundance of (1) those organisms whose burrowing activity left only trace fossils; (2) creatures of many different lineages that evolved hard mineralized tissues; and (3) those soft sea animals that swam or scurried over the surface of the Great Basin seafloor.

The overall consequence of the Cambrian explosions, along with the preceding Ediacaran extinctions, was a complete reorganization of the ocean-floor marine communities. So distinctive is the new set of sea creatures that evolved in the Cambrian that it has been dubbed the Cambrian Fauna, one of three general biotic associations in the Phanerozoic Era (Figure 3.9). The paleontological evidence from the Cambrian rocks of the Great Basin evinces nothing less than a full-scale evolutionary riot that swept across the seafloor more than 500 million years ago!

Cambrian Fossils of the Great Basin

So abundant and varied are Cambrian fossils from the Great Basin that humans have been captivated by them for well over two centuries. Native people of the Pahvant Ute tribe collected Cambrian trilobite fossils in western Utah in the 1800s, fashioning amulets from a dozen or so well-preserved specimens. In the later 1800s, scientists participating in the government surveys of the West first noticed Cambrian fossils, and there has been intensive study of the fossil fauna ever since. The Great Basin region is now regarded as one of the most prolific Cambrian fossil-producing regions in the world. These fossils provide some fascinating windows into one of the most intriguing biotic events to ever affect marine life on Earth.

The earliest Great Basin Cambrian fauna was dominated by deposit and suspension-feeding invertebrate organisms that lived either in the soft sediment or crawled along the muddy seafloor ingesting or straining bits of organic

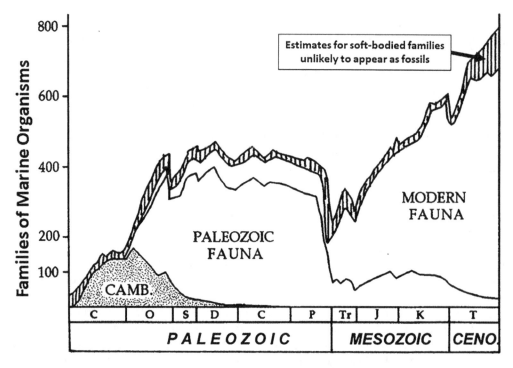

FIGURE 3.9. The three great marine faunas of the Phanerozoic Eon: Paleozoic (541 million–252 million years ago), Mesozoic (252 million–66 million years ago), and Cenozoic (66 million years ago–present.

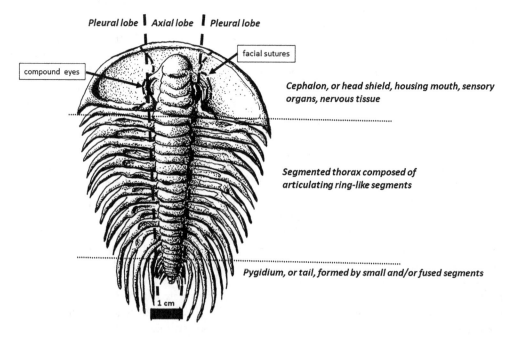

FIGURE 3.10. Generalized features of the trilobite exoskeleton. This trilobite belongs to the genus *Nevadia*, from the early Cambrian of the Great Basin.

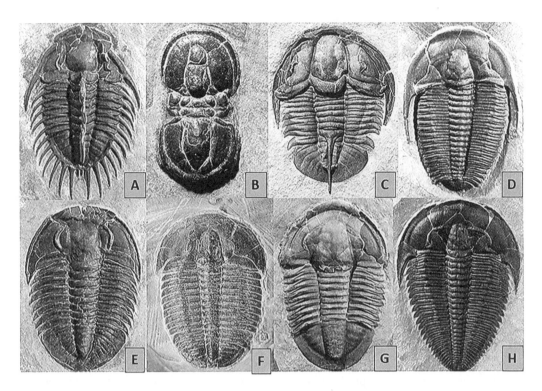

FIGURE 3.11. Common Cambrian trilobites from the Great Basin of western Utah: (A) *Kootenia spencei*, 6 cm long; (B) *Itagnostus interstrictus*, 6 mm long; (C) *Genevievella granulata*, 3 cm long; (D) *Chancia ebdome*, 4 cm long; (E) *Athabaskia bithus*, 5 cm long; (F) *Elrathia kingii*, 2.5 cm long; (G) *Asaphiscus wheeleri*, 6 cm long; (H) *Amecephalus idahoense*, 5 cm long. Credits: A, C, D, E, G, H from Utah Geological Survey Misc. Publication 15-1; B from Parent Géry, Wikimedia Commons, CC_BV-SA 3.0; F from Micha Rieser, Wikimedia Commons CC-BV-SA 3.0

matter in the sediment or from the water. Suspension feeders that filter plankton from the seawater, such as modern bryozoans or many marine worms, were initially less abundant than those obtaining nourishment from the sediment. Among the many fossilized components of the seafloor invertebrate community, the trilobites are overwhelmingly prevalent. In some Great Basin localities, trilobites make up 80–90 percent of the fossils preserved in Cambrian strata; overall, more than 200 different trilobite species have been identified (Figure 3.11).

Trilobites are arthropods (members of the phylum Arthropoda), distantly related to living crustaceans (crabs, shrimp, and lobsters) and insects. The name for this group of extinct arthropods reflects the division of the trilobite body in three lobes: one axial lobe flanked on either side by a lateral (or pleural) lobe (Figure 3.10). Beneath the segmented body of trilobites were many pairs of jointed appendages. One limb in each pair was used for locomotion, while the matching appendage supported gill filaments. The trilobite exoskeleton was a thin carapace composed of calcium phosphate material, like the shells of modern crabs and shrimp. Because the exoskeleton was rigid, a growing trilobite would shed, or molt, the outer shell periodically as new segments were added to the body. Molting was facilitated by a thin crease in the cephalon or "head" region (known as the facial suture, Figure 3.10), which would split apart to allow the naked trilobite to escape its confining sheath. A new and larger exoskeleton was then secreted to protect the heftier body. This growth molting may partially explain the

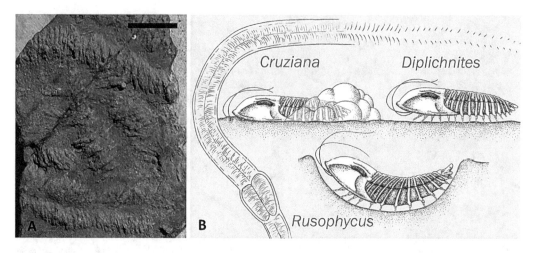

FIGURE 3.12. (A) *Cruziana*, a common trace fossil from Cambrian sediments in the Great Basin; (B) trilobite crawling and burrowing is thought to be related to three different trace fossils, depending on the depth and speed of motion. Drawing from Falconaumanni, Wikimedia Commons license CC BY-SA 3.0.

amazing dominance of trilobite remains among the Cambrian fossils from the Great Basin. A single trilobite could have produced many potentially preservable shells during its life on the Great Basin seafloor: one discarded molt for each body segment, along with the complete shell at death.

The tail, or pygidium, of a trilobite was constructed of small, fused segments, which varied by lineage. In some trilobites, the tails were well developed as broad plates; in others, it was barely noticeable. Some trilobites could flex their bodies enough to cover the unprotected underside with their broad tails. Trilobites had large, crescent-shaped compound eyes that were directed forward and upward, suggesting that they spent most of their time scuttling across the muddy seafloor or perhaps plowing through the mud at depths shallow enough for the eyes to project above the squishy sediment. The distinctive trace fossils made by trilobites (or similar arthropods) moving along the sea bottom are not uncommon in the Cambrian rocks of the Great Basin, even when body remains are absent (Figure 3.12). Generally lacking a hydrodynamic shape and body parts designed for fluid propulsion, most trilobites do not appear to have been active swimmers.

Though they are typically depicted as mobile, bottom-living organisms, there are also some tiny trilobites only a few millimeters long that seem to have adaptations for filling a floating, or planktonic, ecologic niche.

Trilobites appear to have had a variety of feeding habits. Some well-preserved trilobite fossils show long spines near the base of the forward legs near the mouth. These spines might have been used to catch and dismember soft, worm-like organisms from the underlying mud. Others may have been sediment-feeders using their numerous legs to strain organic detritus from the sediment and channeling it forward to the mouth through a medial groove on the underside of the body. Grazing on the carpet of algae and cyanobacteria that covered the seafloor may also have been possible in some groups of trilobites. With so many different types evolving during Cambrian time, it is likely each trilobite lineage evolved specialized anatomy to exploit a variety of ecological niches on the seafloor (Figure 3.11).

In addition to the trilobites, more than 20 other types of arthropods inhabited the Great Basin seafloor during Cambrian time. While we will not attempt a complete description of all the Cambrian arthropods known from the

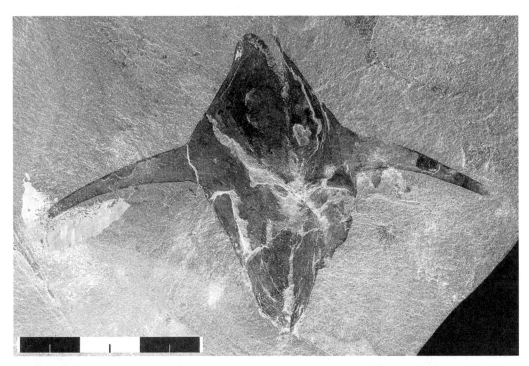

FIGURE 3.13. Carapace of *Pseudoarctolepis*, a phyllocarid arthropod from the Wheeler Shale of the House Range, western Utah. Scale bar divisions in cm. This specimen is housed in the Division of Invertebrate Paleontology, Biodiversity Institute, University of Kansas (KUMIP 144983). Image from the University of Kansas Museum of Invertebrate Paleontology.

Great Basin, we can glimpse the diversity by identifying a few examples of these extraordinary creatures. For example, the phyllocarids were a group of relatively large, primitive crustaceans resembling modern "clam shrimp" or "tadpole shrimp" that belong to the arthropod class Branchiopoda. The ancient phyllocarids, such as *Pseudoarctolepis* (Figure 3.13) and *Canadaspis*, were much larger than their living descendants, but were similar in bearing a hinged carapace that covered most of the body, including the head and jointed legs. *Anomalocaris*, a bizarre predator, was up to 3 feet long and had a circular mouth fed by two large grasping appendages. This odd arthropod propelled itself through the water with wave-like undulations of lateral flaps along both sides of its body (Figure 3.27). *Dicranocaris* was a segmented arthropod about 4 inches long, with a large head and a paddle-like tail, while the smaller *Yohoia* had stalked eyes, a segmented body,

and two antennae-like appendages projecting forward. These examples, and the others illustrated in Figure 3.27, reflect the amazing diversity of arthropods on the Great Basin Cambrian seafloor. The trilobites may be the most iconic group of this phylum, but they coexisted with many relatives within the clan of joint-legged invertebrates.

Other Mineralized Cambrian Fossils from the Great Basin

In addition to the trilobites and related arthropods, Cambrian fossil assemblages from the Great Basin include an intriguing multitude of other kinds of organisms that possessed mineralized exoskeletons (Figure 3.14). All the major phyla of modern marine invertebrates appear in the Cambrian fossil record, including many different types of brachiopods, echinoderms, and sponges. There are also remains of creatures so strange and unlike anything in the modern

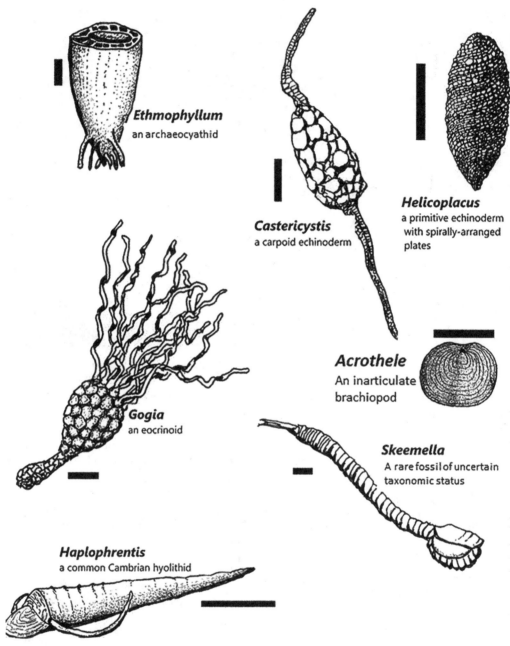

Ethmophyllum
an archaeocyathid

Castericystis
a carpoid echinoderm

Helicoplacus
a primitive echinoderm
with spirally-arranged
plates

Gogia
an eocrinoid

Acrothele
An inarticulate
brachiopod

Skeemella
A rare fossil of uncertain
taxonomic status

Haplophrentis
a common Cambrian hyolithid

FIGURE 3.14. Sketches of non-trilobite Cambrian fossils and organisms from the Great Basin. Scale bar in all sketches = 1 cm.

oceans that their phylogenetic affinities remain mysterious. Let's briefly examine a few of the groups of shelly creatures that shared the Great Basin seafloor with the mob of arthropods.

Many different species of archaeocyathids,

animals like modern sponges (Phylum Porifera), are known from the Cambrian strata of the western Great Basin. The most common forms of the archaeocyathids (Figure 3.14) secreted a cup-shaped exoskeleton up to several inches

FIGURE 3.15. An archaeocyathid reef preserved in the Cambrian Poleta Formation of western Nevada. The light-colored material in the lower part of the outcrop is jumbled archaeocyathid fossils cemented together as part of the reef. The reef was ultimately buried under mud represented by the dark limestone at the top of the photo. Hammer head is 12 cm long.

long, consisting of two perforated cones, one nestled inside the other, and shaped like tiny ice cream cones. The two cones were connected by linear walls (septa) to create a rigid double-cone exoskeleton. The shell was composed of carbonate material and evidently cemented to the seafloor at the small end. With the large, open aperture of the shell facing upward, archaeocyathids appear to have been suspension feeders capable of circulating water through the porous shell to acquire plankton in a manner like many modern sponges. While the cup-like shape is characteristic of most archaeocyathids, some species secreted branching, twig-like, hemispherical, or even pancake-shaped exoskeletons. In many places on the Great Basin seafloor, archaeocyathids grew in such profusion that they formed reef-like masses in shallow water (Figure 3.15). They were the world's first true reef-building organisms and became widespread across the Great Basin Cambrian sea-

floor. Despite their success in the ancient seas, the archaeocyathids were an evanescent group, disappearing forever near the end of Cambrian time. While paleontologists generally agree this group was likely related to the modern sponges, controversy still exists about their precise position on the tree of life.

Several peculiar fossils from the Great Basin are thought to represent early members of the Phylum Echinodermata, which today includes invertebrates with five-fold radial (pentaradial) symmetry such as sea urchins, sea anemones, and starfish. Among the strangest of these early echinoderms are the small (2–3 inches long) carpoids, such as *Castericystis* (Figures 3.14, 3.16). The carpoids belong to an extinct subphylum named Homalozoa, a group of echinoderms so primitive that their slightly flattened body shows no sign of the radial symmetry that characterizes modern echinoderms. In fact, the carpoids show no symmetry at all, a

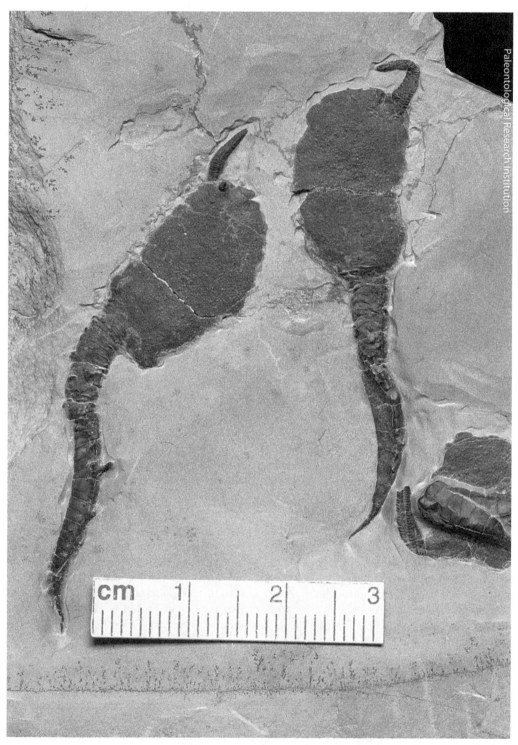

FIGURE 3.16. *Castericystis vali*, a species of carpoid echinoderm from the Cambrian Marjum Formation, Millard County, Utah. Specimen from the Paleontological Research Institution, Ithaca, NY.

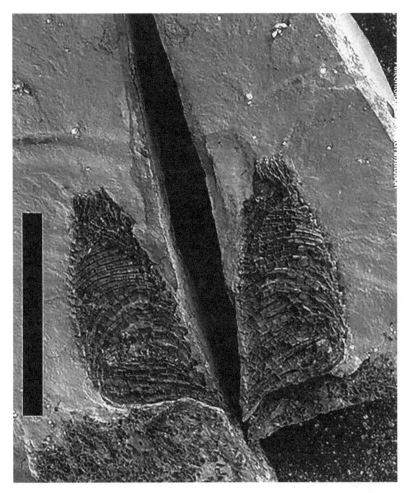

FIGURE 3.17. *Helicoplacus* from the Poleta Formation of the White-Inyo Mountains, eastern California. Scale bar = 10 cm. Specimen from the Paleontological Research Institution, Ithaca, NY.

fact that makes them difficult to associate with any living group of invertebrates with certainty. However, like most living echinoderms, the carpoids armored their body with calcite plates and appear to have food grooves made of smaller plates. Some paleontologists have suggested that the carpoids may have had gill openings and an internal water vascular system like that of modern sea urchins. Well-preserved carpoid fossils commonly have a "tail" of some unknown function, and some forms possessed a smaller single "arm" that may have been used in feeding or locomotion. Both the tails and the arms are made of small calcite plates. Because they are so dissimilar from their living relatives, little is known with certainty about carpoid biology. However, it has been suggested that they possessed a nervous system sufficiently elaborate to make them distantly related to chordates. Accordingly, the term "calcichordate" has been used by some scientists to the express this possible relationship.

Another group of odd echinoderms in the Cambrian fauna are the helicoplacoids (Figures 3.14, 3.17), whose football-shaped exoskeletons were constructed of a mosaic of plates arranged in a spiral fashion. There are three food grooves that spiral around the spindle-shaped body.

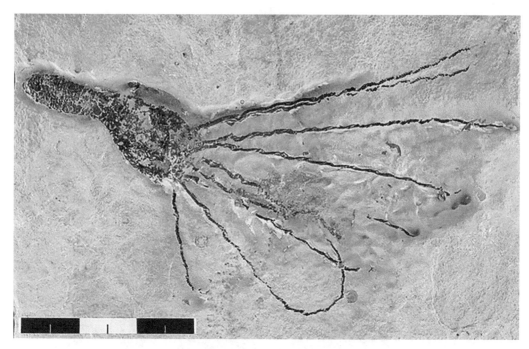

FIGURE 3.18. *Gogia spiralis*, an eocrinoid from the Wheeler Shale of western Utah. This specimen is housed in the Division of Invertebrate Paleontology, Biodiversity Institute, University of Kansas (KUMIP 314064). Image from the University of Kansas Museum of Invertebrate Paleontology.

These early echinoderms probably were suspension feeders, either standing upright on the seafloor or partially buried in the soft sediment. The mouth was located along the food grooves and may have been positioned at the top, or along the side, of the helical body. Complete helicoplacoid specimens may be as long as about 3 inches (8 cm) but are extremely rare. The Early Cambrian Poleta Formation of the White-Inyo Mountains of the western Great Basin is one of the very few rock units in the world where nearly complete helicoplacoid exoskeletons have been preserved as fossils. In this location, they occur in shale representing mud that smothered entire groups of helicoplacoids living in relatively deep and calm water just offshore (westward) from the shallow shelf. The rapid burial may have been triggered by earthquakes or violent storms that dislodged soft mud from the edge of the shelf, sending billowing clouds of fine sediment into the deeper basin. In any case, the sudden entombment of the helicoplacoids limited postmortem decay

and disaggregation of their skeletons. The unusual abundance and preservation of helicoplacoid fossils in the western Great Basin may thus reflect events and circumstances unique to the deep-water basins on the seafloor west of the shelf edge.

Another group of Cambrian echinoderms on the Great Basin seafloor were the eocrinoids (Figures 3.14, 3.18). The eocrinoids had a globular, cup-shaped body (theca) composed of calcite plates. The mouth was positioned at the opening of the cup and faced upward. A varying number of flexible brachioles, or arms, extended upward into the water from the mouth and were apparently used in suspension feeding. At the base of the body was a short expandable stalk that functioned as a holdfast, anchoring the eocrinoid in the mud of the seafloor or cementing it to other skeletal material. These features are like the modern echinoderm group known as crinoids, or "sea lilies." However, the Cambrian forms are smaller and much more primitive than modern crinoids and only distantly related. Eocrinoids were the most common of the Cambrian echinoderms but disappeared in the Silurian Period when several other groups of more advanced, stalked echinoderms became abundant and widespread on the Great Basin seafloor. The genus *Gogia* is by far the most common Cambrian eocrinoid in the Great Basin, with many species thus far identified from the region.

There are also many different sponges in the Great Basin Cambrian faunas. Sponges are simple metazoans that comprise the Phylum Porifera in the modern world. These primitive invertebrates are comprised mostly of soft tissues with specialized cells to circulate water through the porous body, to capture organic matter as food, to process the food, for respiration, and to support the body. Skeletal material in most sponges is limited to small spicules of silica, calcite, or fibrous organic matter woven through the body to provide support. Postmortem decay of sponges typically destroys the soft tissues, leaving only the tiny, mineralized spicules as potential fossils. Many kinds of sponge spicules have been discovered in Cambrian rocks of the Great Basin, indicating that a variety of different poriferans inhabited the seafloor. While they are rare, nearly complete and well-preserved sponge fossils have also been collected from fine-grained rock units such as the Wheeler shale and Spence shale of Utah (Figure 3.19). These exceptional specimens demonstrate that Cambrian sponges were, for the most part, similar in shape and construction to the basket-like or encrusting forms of their modern descendants.

The Cambrian fossil fauna of the Great Basin also includes several representatives of both the phyla Mollusca and Brachiopoda. Modern molluscs include clams, oysters, snails, slugs, squid, and other less-common forms. Brachiopods are known as lampshells and are now less common than molluscs, but they were the dominant shellfish of the Paleozoic Era. Though the molluscs and brachiopods have quite different soft anatomy, both groups include species that secrete calcareous shells to protect their soft bodies. The durable shells are likely to be pre-

FIGURE 3.19. A nearly complete specimen of *Vauxia magna*, a middle Cambrian sponge preserved in the Spence shale of the Wellsville Mountains, Utah. Scale bar = 5 cm. This specimen is housed in the Division of Invertebrate Paleontology, Biodiversity Institute, University of Kansas (KUMIP 111763). Image from the University of Kansas Museum of Invertebrate Paleontology.

served as fossils, so it is not surprising that these phyla appear in the Cambrian explosion. However, the earliest molluscs were generally very small, some with features quite unlike any of their living relatives. This can make the identification of some Cambrian shells as molluscs a bit controversial.

For example, one of the most distinctive groups of Cambrian organisms are the hyolithids such as *Haplophrentis* (Figures 3.14, 3.20). These animals possessed a three-sided conical shell, about an inch long, with a semicircular lid (operculum) over the open end. Two lateral blades or spines projected outward

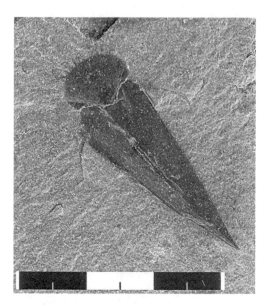

FIGURE 3.20. A well-preserved fossil of *Haplophrentis reesei*, a hyolithid from the Cambrian Spence shale of the Wellsville Mountains, Utah. Scale bar divisions in cm. This specimen is housed in the Division of Invertebrate Paleontology, Biodiversity Institute, University of Kansas (KUMIP 204340). Image from the University of Kansas Museum of Invertebrate Paleontology.

and backward from the cone's open end. Traditionally, hyolithids were thought to be primitive molluscs, but recent studies of well-preserved specimens from Utah and British Columbia have led scientists to associate them with the brachiopods. They are now thought to have been suspension feeders, lying on the ocean floor while raising the operculum-bearing end up with the lateral spines. A comb-like organ known as the lophophore evidently extended from the open operculum to strain food particles and perhaps support gill filaments. The lophophore is a key feature of the brachiopods, and its discovery in hyolithids from the Great Basin and nearby regions has led some paleontologists to reclassify these odd creatures from the phylum Mollusca to the phylum Brachiopoda.

Many other less ambiguous brachiopod and molluscan shells have been identified from

Cambrian strata in the Great Basin, sometimes in great abundance. However, the shells are typically small (only a few millimeters in size) and can be easily overlooked. Most of the Cambrian brachiopod fossils of the Great Basin are small, semicircular phosphatic shells (Figure 3.14). Though brachiopods, like modern clams, are bivalves (possessing two shells, or valves), the fossils of Cambrian forms are generally found as single shells, unconnected by any type of tooth-and-socket hinge seen in later marine bivalves. Brachiopods without a strong connection between the two shells comprise a group known as the "inarticulates," one of two broad subdivisions of the phylum. Cambrian molluscs from the Great Basin include cap-shaped shells similar in form to modern limpets, along with coiled shells that look like tiny snails. Both limpets and snails are gastropods, a class of molluscs with one variably coiled shell. However, compared to modern gastropods, the Cambrian forms are only a few millimeters in size. That paleontologists have only identified a few genera of Cambrian molluscs may reflect the ease of overlooking such tiny and simple shells among the multitude of much larger and gaudier fossils preserved in the strata of the Great Basin.

These groups of shell-bearing organisms all burst into abundance at about the same time and are important elements of the Cambrian explosion and the Cambrian Fauna (Figure 3.21). Some of the major phyla, such as sponges (Porifera) and jellyfish (Cnidaria) may have evolutionary roots that extend back into the Neoproterozoic Era, but even these organisms become suddenly more abundant and diverse in Early Cambrian time. But however dramatic the increase of preserved fossils in Cambrian strata of the Great Basin may seem, it is just a hint of the real depth of the evolutionary outburst on the seafloor.

Due to some very unusual conditions of preservation, the remains of many groups of Cambrian soft-bodied organisms have been discovered in recent years in the Great Basin. These fossils may represent creatures that may have been as dominant in the Cambrian fauna as those with mineralized tissues. Without

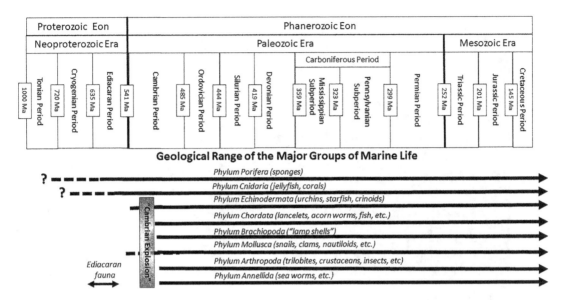

FIGURE 3.21. Geological ranges of the major phyla of marine invertebrates. Note that most of these groups first appear as abundant fossils simultaneously in the Cambrian Period.

durable skeletal material, such soft-tissue organisms generally decay completely on the seafloor after death under the combined assault of scavenging creatures, microbial decomposition, and physical destruction from currents and waves. But on the Cambrian Great Basin seafloor, local conditions allowed for the exquisite preservation of some soft-bodied organisms as impressions and carbon-rich films in fine mud that later lithified into shale or silty limestone. These fossils are sometimes spectacular in detail and have provided a new window into the richness of the Cambrian fauna. What were these creatures, how were they preserved, and what do their remains tell us about conditions on the Cambrian Great Basin seafloor?

Cambrian Lagerstätten of the Great Basin

In 1970, German paleontologist Adolph Seilacher first used the term *Lagerstätte* (pl. *Lagerstätten*) for an exceptionally rich and/or extraordinarily well-preserved accumulation of fossils. Later work by Seilacher and his colleagues recognized two types of Lagerstätten: (1) *Konzentrat-Lagerstätten* (concentration Lagerstätten) and (2) *Konservat-Lagerstätten*

(conservation Lagerstätten). In the former, the fossil accumulation is distinguished by unusual abundance of hard parts, such as a bone bed or a shell bed. In Konservat-Lagerstätten, the accumulation is unique by virtue of exceptional preservation, not necessarily the abundance, of the fossil remains. In some cases, a single Lagerstätte may qualify as both types. Such a fossil accumulation is a treasure to paleontologists, who normally acquire only fragmentary and biased data on prehistoric life from the rock record.

Lagerstätten are known from all over the world and occur in sedimentary rocks of all ages. However, western North America is home to an unusual number of Cambrian Lagerstätten, suggesting that there must have been something unique happening on the Great Basin seafloor to preserve fossils in such spectacular abundance and quality in several different places. In the Great Basin region, Cambrian Lagerstätten have been discovered in the Wellsville Mountains of northern Utah in the Spence shale member of the Langston Formation. In the House Range and Drum Mountains of western Utah, Lagerstätten occur in the Wheeler Shale (Figure 3.23) and Weeks, Marjum, and Pierson Cove Formations. Similar fossil accumulations

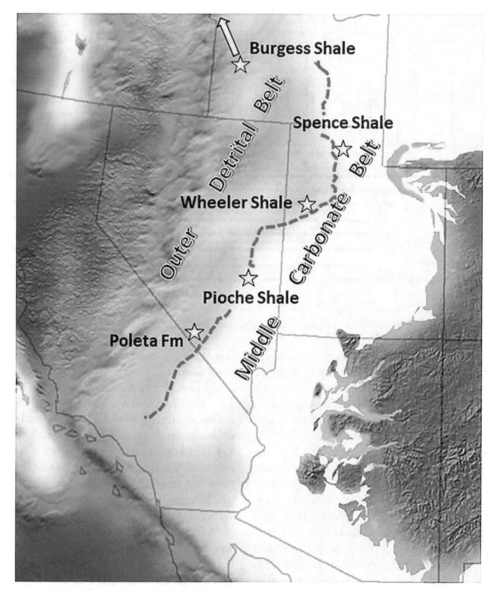

FIGURE 3.22. Approximate centers of deposition (stars) for the rock units that host Cambrian Lagerstätten (exceptionally rich and/or extraordinarily well-preserved fossil accumulations) in the Great Basin. The Burgess Shale localities are also along this trend, but much farther north in British Columbia. Base map is a paleogeographic reconstruction from R. Blakey.

have been documented in the Pioche Shale of the Chief Range in eastern Nevada, and in the Poleta Formation of western Nevada (Figure 3.22). To the north of the Great Basin, but still related to the submerged passive margin of Laurentia, the Burgess Shale in eastern British Columbia houses what might be the most renowned Lagerstätte in the world. In these locations, the extraordinary fossils of Middle Cambrian age are preserved in dark shales that formed at about the same time and in similar depositional settings across the Great Basin.

While each Cambrian Lagerstätte in the Great Basin has its own preservational history

FIGURE 3.23. Outcrop of the Wheeler Shale in the House Range of western Utah. Note the dark color and thin, undisturbed bedding of this calcareous shale. The Cambrian Lagerstätten of the Great Basin are commonly preserved in such sediments.

and faunal composition, there are some similarities between them that suggest similar circumstances were involved in the origin of these exceptional fossil caches. First, the fossils are generally preserved in fine-grained calcareous shale and clay-rich limestone that originated as soft mud on the seafloor, just offshore from the shallow carbonate bank (Figure 3.22). The shales are commonly thin-bedded, carbon-rich, and contain tiny, dispersed crystals of pyrite (FeS_2), a mineral that forms in the absence of oxygen. These features suggest that the sediment accumulated in stagnant, oxygen-deficient (anoxic or dysoxic), poorly circulating bottom waters. Furthermore, the Lagerstätten are commonly preserved in sediments that have thin, even layering, undisturbed by burrowing organisms. This indicates that the minimal oxygen levels probably limited populations of scavenging and burrowing organisms that might have otherwise destroyed organic remains. Detailed studies of the strata enclosing most of the Great Basin Cambrian Lagerstätten suggest that the fine mud was deposited in discrete events after periods of nondeposition. In these cases, it appears that storm events may have periodically dislodged fine mud and clay on the outer edge of the carbonate bank, sending clouds of suspended sediment onto the deeper ocean floor to smother and bury the organisms it transported (Figure 3.24).

Finally, the fine-grained texture and calcareous composition of the sediments preserving the Lagerstätten fossils likely inhibited the flow of fluids through the mud after deposition, preventing extensive postburial decay and chemical degradation of buried organic material. The result of these factors operating together in several different spots on the Great Basin seafloor produced some of the most dazzling concentrations of beautifully preserved Cambrian fossils in the world (Figure 3.25).

The remarkable preservation of fossils in the Cambrian Lagerstätten of western North

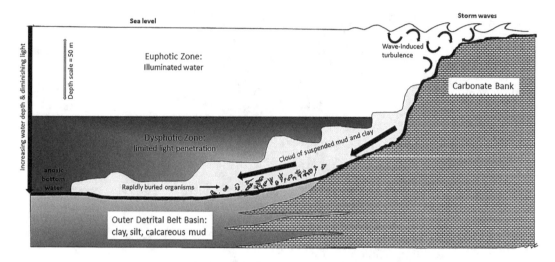

FIGURE 3.24. Origin of Cambrian Lagerstätten of the Great Basin. Masses of fine sediment dislodged by storms from the edge of the carbonate bank may have transported and smothered organisms in deeper water offshore where anoxic conditions limited decomposition.

FIGURE 3.25. Exceptional preservation of the trilobite *Elrathia kingii* from the Wheeler Shale Lagerstätte of the House Range, Utah. Such preservation of complete specimens exhibiting even the finest details of the exoskeleton is extraordinarily rare.

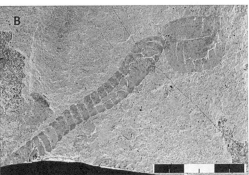

FIGURE 3.26. Examples of preservation of soft-bodied organisms in Great Basin Lagerstätten: (A) mouth appendage of *Anomalocaris* from the Pioche Shale, Nevada (scale bar in mm); (B) *Skeemella clavula*, an odd arthropod from the Pierson Cove Formation of the Drum Mountains, Utah (scale bar in cm). Both specimens housed in the Division of Invertebrate Paleontology, Biodiversity Institute, University of Kansas (A: KUMIP 29375; B: KUMIP 310501).

America reveals the presence of many soft-bodied organisms that would normally vanish from the geological record without a trace. For example, the very strange arthropod predator *Anomalocaris* was unknown until its soft body parts were discovered and analyzed by paleontologists (Figures 3.26A, 3.27). The body of *Anomalocaris* was comprised mostly of soft tissues, but the mouth parts were at least partially mineralized. This menacing predator was up to 3 feet long and had a circular mouth bordered by two large, spiked appendages. It was a free-swimming marauder that swept over the seafloor by undulating the lateral lobes along its body (Figure 3.27).

Another example of an oddity from the Great Basin Lagerstätten is *Skeemella clavula*, first discovered in the Pierson Cove Formation of Utah (Figure 3.26B). This strange creature had a worm-like, segmented body a few inches long that terminated in a forked tail. The front end of *Skeemella* was expanded in a "head" comprised of several pouch-like components covered by a carapace. The segmented body of *Skeemella* is similar to arthropods (except for the absence of legs), while the complex head may represent an early chordate. *Skeemella* was no doubt a swimming animal and may have fed either on plankton in the water column or organic mat-

ter in the sediment. Other aspects of its biology remain so uncertain that most scientists classify it as a member of the Vetulicolia, a phylum or subphylum of problematical extinct metazoans thought to be linked to the earliest chordates.

The fossils from Cambrian rocks in the Great Basin provide invaluable information on the character of the Cambrian fauna populating the Great Basin seafloor. This unique assemblage of marine organisms was incredibly diverse and included the earliest members of modern phyla, as well as many forms that ultimately vanished near the end of Cambrian time (Figure 3.27).

The explosive development of the Cambrian fauna across the Great Basin seafloor was the consequence of numerous factors, including the shift from microbial mats to intensely burrowed substrates, progressive oxygenation of bottom sediments, adaptation to specific ecological niches in an expanding marine food web, and possible changes in ocean chemistry. The combination of these factors was unique to the Cambrian Period; never again would the ocean floor experience such an extraordinary burst in the diversity of life. However, near the end of Cambrian time, after 50 million years of frenzied evolution, things changed on the Great Basin seafloor.

FIGURE 3.27. Reconstruction of the Cambrian fauna of Laurentia based on fossils from the Burgess Shale, Wheeler Shale, Pioche Shale, and Spence Shale Lagerstätten. Original art by Carel Brest van Kempen. Reproduced with permission of the artist.

The Cambrian Extinctions and Demise of the Cambrian Fauna

Beginning around 500 million years ago, several waves of extinction swept through the oceans of the world, the last one peaking at about 488 million years ago. The trilobites, sponges, and brachiopods were the main victims of these extinction events, but many of the soft-bodied organisms unique to the Cambrian Fauna also disappeared or suffered serious decline. Geologists and paleontologists are still debating the causes of the Late Cambrian extinctions, but there are some indications of multiple events that could have produced the biotic stresses that eradicated many lineages of marine invertebrates. There is, for example, evidence in Early Ordovician sediments (about 460 million years old) of global glaciation, suggesting

that the warm climate of the Cambrian Period was rapidly deteriorating about the time the extinctions began. Lower water temperatures could have affected many of the organisms of the Cambrian fauna, most of which evolved in, and were well adapted to, warm seas.

Also, based on sulfur isotope signatures in the rocks, it has been suggested that oxygen levels in the seas may have declined near the end of Cambrian time, resulting in widespread transitory anoxia on the seafloor. In addition, advancing glaciers on land drew water from the oceans, causing a decline in sea level and regression. The temporary withdrawal of seas would have eliminated at least some of the shallow marine habitat that harbored the Cambrian fauna. However, the regression of the Late Cambrian seas in western North America was

FIGURE 3.28. Key to the organisms depicted in Figure 3.27: (1) stromatolites; (2) *Leptomitus*, a sponge; (3) *Vauxia*, a sponge; (4) *Billingsella*, a brachiopod; (5) *Hallucigenia*, an arthropod; (6) *Aysheaia*, a predatory arthropod; (7) *Anomalocaris*, a huge swimming arthropod predator; (8) *Opabinia*, a five-eyed predator; (9) *Lejopyge*, a tiny trilobite; (10) *Olenoides*, a trilobite; (11) *Asaphiscus*, a trilobite; (12) *Elrathia*, a very common trilobite; (13) *Modocia*, a trilobite; (14) *Naraoia*, a putative trilobite; (15) *Habelia*, a bottom-living invertebrate; (16) *Burgessia*, an extinct arthropod similar to modern horseshoe crabs; (17) an unnamed, tadpole-like arthropod; (18) *Odaraia*, a primitive crustacean; (19) *Sarotrocerus*, an enigmatic back-swimmer; (20) *Pseudoarctolepis* and (21) *Canadaspis*, both phyllocarid arthropods; (22) *Marella*, an arthropod; (23) *Branchiocaris*, an arthropod; (24) *Ottoia*, a predatory worm; (25) *Hyolithes*, a early brachiopod(?); (26) *Canadia*, a segmented worm; (27) *Gogia*, an eocrinoid; (28) *Pikaia*, a worm-like cephalochordate; (29) *Wiwaxia*, an armored invertebrate; (30) *Dinomischus*, unknown group; (31) *Amiskwia*, unknown group. Original art by Carel Brest van Kempen. Reproduced with permission of the artist.

temporary and was soon followed by renewed transgression related to the peak of the Sauk sequence.

The Late Cambrian extinctions occurred at a time when carbon, oxygen, and sulfur isotope data suggest that chemical cycling in the sea was repeatedly disrupted. Researchers have discovered that carbon-13 becomes slightly more enriched relative to carbon-12 in the Cambrian sediment that accumulated as the extinctions began. This geochemical hiccup is now called the Steptoean (a subdivision of the Late Cambrian rock record) Positive Carbon Isotope Event, or SPICE. The SPICE anomaly occurs in several different rock units in the Great Basin. It persisted for about 4 million years and appears to be a global event. One interpretation of this discovery is that more organic carbon (which contains higher levels of carbon-12 than seawater) was being buried on the ocean floor. This could have been because organisms were dying from anoxic conditions, and the carbon in their tissues rained to the ocean floor and remained there without decomposition. It could also suggest a sudden bloom of plankton that extracted carbon-12 from the seawater and carried it to the ocean floor after death.

Oxygen isotope data from Late Cambrian rocks also document a cooling of the oceans of western Laurentia that may have resulted from climate shifts or from enhanced upwelling of bottom water. Positive shifts in sulfur-34 also occur in the Late Cambrian in a manner consistent with the increased formation of pyrite (FeS_2) in the anoxic mud of the seafloor. The causes and significance of these geochemical anomalies are still being debated, but their nearly simultaneous occurrence in rocks that formed during the Cambrian extinctions indicate that the normal cycling of material through the marine ecosystem was stunted by rapid environmental changes as the extinction commenced.

Whatever the causes of the Late Cambrian extinctions, this biotic crisis forever changed the character of seafloor life. Many of the iconic members of the Cambrian fauna, such as the trilobites, archaeocyathids, and exotic arthropods like *Anomalocaris*, either vanish from the fossil record or withered in abundance and diversity during the upheaval. Yet there were some survivors. About 25 percent of the Cambrian trilobite families struggled through the extinction, accompanied by a few groups of brachiopods, molluscs, echinoderms, and even chordates. As the Cambrian fauna wilted, these survivors gave rise to new groups of descendants in the later periods of the Paleozoic Era.

The faunal turnover at the end of the Cambrian Period ultimately led to a whole new array of marine organisms on the seafloor that remained stable for more than 200 million years. This new congregation of organisms comprises what is known as the Paleozoic fauna (Figure 3.9), a diverse set of fascinating sea creatures that we will continue to encounter for the duration of the Great Basin seafloor.

4

The Ordovician Overhaul

When the sun rose on the first day of the Ordovician Period some 485 million years ago, nothing much changed on the Great Basin seafloor, and there was no indication that a new chapter in the natural history of the region had begun. Laurentia was almost entirely submerged by warm and luminous tropical seas, as the Sauk transgression was reaching its peak. At that point, sea level may have been nearly 2,000 feet higher than it is today, and only a few small areas of Laurentia remained unsubmerged. The northwestern part of the ancient continent, where the modern Great Basin was situated, was located a few degrees south of the equator (Figure 4.1), and soft trade winds swept over the shallow seas generating currents that circulated oxygen and nutrients across the continental shelf. The survivors of the Late Cambrian extinctions, including trilobites, echinoderms, and brachiopods, populated the shimmering seafloor beneath the waves, while worm-like creatures churned through the underlying mud and ooze.

In this tranquil setting, there would have been no hint of the sweeping changes that were soon to affect the Great Basin seafloor. Globally, however, new plate tectonic interactions were developing that ultimately led to dramatic changes in world climate, geography, and ocean chemistry. The Great Basin seafloor was about to be dramatically reshaped by a combination of interconnected biologic, geologic, and climatic events. It was in the Ordovician Period that a long series of geologic disturbances linked to the convergence of tectonic plates began to affect the eastern margin of Laurentia. In what is now New England and the north Atlantic coastal plain, the oceanic crust attached to the ancient continent was consumed to the east beneath a chain of offshore volcanic islands. As the volcanic island arc migrated west toward land, the narrow seaway gradually closed.

By Late Ordovician time, the volcanic chain collided with the continent, crushing older rocks, and raising a major mountain system in what geologists have named the Taconic orogeny. This mountain-building event takes its name from the Taconic Mountains in eastern New York and New England. It was the first of several major orogenic events that would deform and lift the bedrock of the modern Appalachian Mountains. Elsewhere, most notably in the southern hemisphere, other landmasses were converging toward a giant continent known as Gondwana. The piecemeal construction of Gondwana had begun much earlier, but accretion in Ordovician time enlarged this landmass to supercontinent status.

The oceanic realm in Ordovician time was dominated by a single global ocean known as the Panthalassa (Figure 4.1), with narrow seaways such as the Iapetus Sea bordering eastern Laurentia and separating the various landmasses. Ultimately, as we will soon see, the assembly of Gondwana and its movement toward the south pole, had profound effects on the global climate.

Middle Ordovician
465 Ma

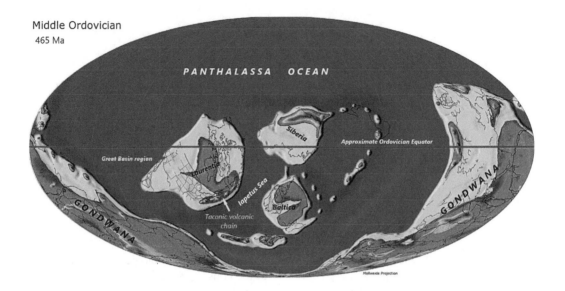

FIGURE 4.1. Global paleogreography of the Middle Ordovician Period, approximately 465 Ma. Base map from Scotese and Wright (2018), PALEOMAP Project (https://www.earthbyte.org /paleodem-resource).

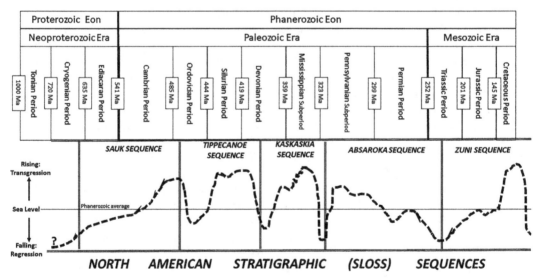

FIGURE 4.2. Long-term eustatic variations and Sloss sequences of the Paleozoic Era. Sea level fluctuations are the result of combined tectonic, climatic, and oceanographic events, while the Sloss sequences are the sedimentary responses to these changes.

Sediment eroded from the rising Taconic mountains spread over much of Laurentia, covering the thick limestone sequences formed during the Sauk transgression with sand, silt, and mud. This mountain-building episode also buckled the land farther inland, causing the seas to withdraw from areas previously submerged. The regression exposed tracts of Late Cambrian and Early Ordovician sediments to erosion. Regional unconformities marking this period of exposure are common in the Ordovician rock record across North America. However, in Middle Ordovician time, this erosion surface would be submerged again when sea level rose during another great inundation of Laurentia known as the Tippecanoe sequence (Figure 4.2). The Tippecanoe sequence, along with the preceding Sauk sequence and the subsequent Kaskaskia sequence, was named by geologist Lawrence Sloss in 1963. The rocks recording these episodes of eustatic, or sea-level fluctuation are therefore commonly referred to as "Sloss sequences." As Sloss defined it, a sequence is a set of rocks that formed from sediment accumulated during a period of prolonged inundation of ancient North America. A sequence is bounded above and below by major unconformities that reflect times of exposure and erosion when sea level was relatively low prior to and after the continental flooding. Across North America, the Tippecanoe sequence consists of rocks deposited after the Middle Ordovician erosion surface was submerged by advancing seas until Early Devonian time, when a major drop in relative sea level once again exposed vast tracts of inland Laurentia to erosion.

Before the volcanic activity in the Taconic arc subsided, great clouds of ash settled into the seas and on land to produce layers of bentonite (clay that forms primarily from weathered volcanic ash), another unique attribute of the Ordovician strata of central North America. The Taconic orogeny did not directly affect the Great Basin seafloor, but it did play a role in reversing the Sauk transgression and helped to initiate a continent-wide regression of seas that ultimately reached the western edge of Laurentia in Ordovician time. As the seas withdrew from the interior of Laurentia, the water covering the Great Basin seafloor became extremely shallow and eventually drained completely away in some areas before the Tippecanoe transgression resubmerged the region. The Ordovician rock record in Utah and Nevada, as we will see, clearly reflects these global eustatic events.

During Ordovician time, there were also other influences, superimposed on the tectonic events, that promoted even more sea level insta-

bility. The climate of western Laurentia in the Early Ordovician Period was, as it had been in Cambrian time, warm, wet, and tropical. Near the end of the Ordovician, in a narrow slice of time known as the Hirnantian age (445–443 Ma), the Earth experienced a major glaciation. During this ancient ice age, glaciers grew primarily in the southern hemisphere, where the enormous supercontinent of Gondwana had moved over the south pole. In fact, the motion of Gondwana into this polar position may have helped trigger the Hirnantian glaciation by reflecting more solar energy into space than would energy-absorbing seawater. While no glacial ice formed along the mostly submerged edges of Laurentia during the Hirnantian glaciation, conditions on the Great Basin seafloor were strongly influenced by the Late Ordovician climate deterioration. Studies of carbon isotopes in the rocks of the Great Basin and elsewhere suggest that the temperature of the tropical seas fell from more than about 40°C at the beginning of the Ordovician Period to less than 25°C during the Hirnantian ice age. In response to the growth of massive glaciers in Gondwana, global sea level fell by 150–300 feet. In Laurentia, this decline was superimposed on the eustatic changes associated with the tectonic activity in progress along the continent's eastern margin. As the seas covering western Laurentia fell and rose, the Great Basin seafloor was exposed several times. The erosion that occurred during each period of exposure resulted in the numerous unconformities and cavities (described later in this chapter) in the rock record.

If these Ordovician events seem to suggest a somewhat chaotic setting across the Great Basin seafloor, there is even more turmoil to ponder. About 20 years ago, scientists discovered that Middle Ordovician limestone in southern Sweden contained an unusual number of small (1–4 cm) meteorites. The composition of minerals such as chromite and spinel from the meteorites, along with the isotopic composition of the rare metal osmium, indicate that they came from the same large extraterrestrial body. Specifically, the meteorites are of a type known as

L chondrites, thought to have originated from asteroids that collided and broke apart about 470 million years ago. Within a few million years of their violent origin, many of these fragments drifted toward earth and caused a global storm of meteors when they fell through the Ordovician atmosphere. After their initial discovery in Sweden, fossil meteorites and chromite grains from them have also been found in Middle Ordovician rocks in China, Russia, Scotland, Argentina, and North America. Several of the meteorites appear to have been large enough to have excavated craters up to several miles wide when they struck the surface. Four of these craters, now deeply buried in the subsurface between Oklahoma and the Great Lakes, were revealed in the 1990s when geologists searching for oil examined geophysical images of the deep subsurface of central North America.

How many other impacts occurred during the Ordovician meteor storm is unknown, but it may have been in the hundreds. Overall, scientists estimate that the rate of meteor bombardment may have been 100 times more intense than it is today, lasting a period of several million years. Interestingly, not all the material formed during the asteroid collision has yet fallen to earth. Some of it still drifts within the asteroid belt, while other fragments spiraled sunward but did not intersect the Earth's orbit. In fact, about a third of all meteorites falling to earth today are of the L chondrite type. These space rocks are still arriving here, some 470 million years after they were created.

The consequences of the rain of cosmic debris during the Middle Ordovician meteor storm are still being debated by geologists and paleontologists. Meteoritic dust lingering in the atmosphere and orbiting near the earth may have reflected enough of the sun's energy to initiate a climatic cooling trend. Minerals entering the seas from the incoming meteorites and their ashes may have provided nutrients for plankton or, conversely, may have introduced toxic metals that harmed planktonic organisms. The explosive excavation of an unknown number of impact craters may have created clouds of dust and rock fragments that fell into the seas,

reducing water clarity. While scientists have yet to precisely reconstruct the effects of the Ordovician meteor storm in the global ocean, the event certainly contributed to the wave of turmoil that swept across the Great Basin seafloor. The Ordovician rock record of western North America clearly reflects a juddering sea level and evinces a simultaneous shift in the character of sea life.

Ordovician Rocks of the Great Basin

The sediments that accumulated on the rifted margin of Laurentia during Ordovician time followed the same general pattern established in the Ediacaran Period. Nearest to shore, in what is now Utah and eastern Nevada, shallow marine deposits continued to settle onto the continental shelf. Carbonate rocks (limestone and dolomite), with lesser amounts of shale and sandstone, dominate the shelf strata and form a generally west-thickening wedge known as a miogeocline (Figure 4.3). The most widespread shallow marine rock unit is the Pogonip Group, comprised of several different formations in the central and eastern Great Basin. Farther offshore, in western Nevada and eastern California, fine silt and clay accumulated in deeper water beyond the edge of the continental shelf. This area of deep-water sedimentation is traditionally known as the eugeocline, the deepest part of the Great Basin seafloor in Ordovician time. Eugeoclinal rocks are generally comprised of black shales and associated chert and submarine volcanic rocks. Such rocks dominate the Vinnini and Valmy Formations of western and central Nevada. As we will see in Chapter 6, the Ordovician deep marine sediments were thrust eastward over the edge of the continental shelf in later Paleozoic time. Crumpled and shattered masses of these deep marine rocks are now widespread in central Nevada as deformed strata in the upper plate of regional thrust faults. Estimating the eastward displacement on the thrust faults allows geologists to determine the approximate location of the deep basins where the eugeoclinal sediment originally accumulated. This deep ocean floor appears to have existed west of central Nevada, extending at

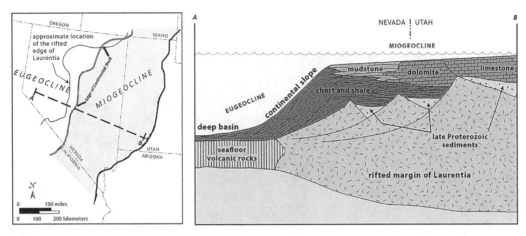

FIGURE 4.3. Map view (*right*) and cross-section (*left*) of the miogeocline and eugeocline in the Great Basin region during early Paleozoic time.

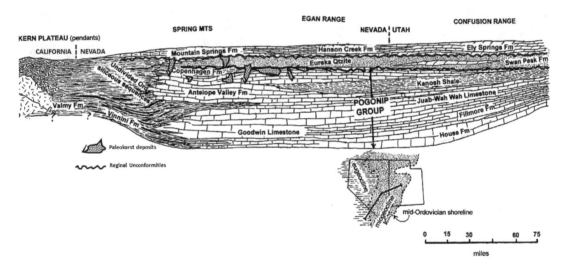

FIGURE 4.4. Cross-section of the stratigraphy of Ordovician rocks in the Great Basin region.

least as far as the modern Sierra Nevada in California, beyond which no Ordovician rocks are preserved.

The Ordovician Miogeocline: Fitful Seas and a Flood of Sand

Because the water depth over the miogeocline was minimal, the eustatic oscillations of Ordovician time are most clearly reflected in the rock record of the shallow seafloor in the eastern Great Basin. As sea level rose and fell in rhythm with Ordovician climatic and tectonic cycles, the abyssal ocean floor of the eugeocline

remained deeply submerged, while the miogeocline was exposed and resubmerged numerous times. The rock record of the miogeocline clearly documents a pattern of recurring exposure and drowning throughout Ordovician time. Many exposures of the miogeoclinal strata are also richly fossiliferous, allowing us to recognize some fascinating patterns in the evolution of life across the inner Great Basin seafloor.

Miogeoclinal strata in the eastern Great Basin can be broadly subdivided into three components. The Early Ordovician rocks are mostly limestones and shales deposited in very

FIGURE 4.5. Ordovician rock units exposed on the west flank of the Arrow Canyon Range, southern Nevada.

shallow water. The Pogonip Group, which consists of a variable number of formations across the Great Basin, is the iconic rock unit of the Early Ordovician miogeocline (Figures 4.4, 4.5). Ranging in thickness from about 1,000 feet to 3,500 feet, this sequence of limestone and shale has many indications of extremely shallow water and exposure related to several significant drops in sea level during the Early and Middle Ordovician. Irregular erosion surfaces produced during periods of exposure are preserved in the rock record as unconformities, some of which are of regional extent. Such widespread unconformities (Figure 4.5) provide evidence that the falling seas exposed vast areas of the miogeocline. When the shallow shelf was submerged, the water depth was probably less than about 150 feet, even far offshore.

In such shallow water, strong currents driven by tidal surges, prevailing winds, and storms would rip across the miogeocline repeatedly. When such storms swept over the miogeocline of the Great Basin, furious waves and bottom currents would shred the semiconsolidated carbonate mud on the Great Basin seafloor to produce the flat pebbles preserved in the conglomerate and the uneven and irregular layering commonly seen in formations comprising the Pogonip Group. These accumulated deposits are now recognized as a distinctive rock composed of the lithified rubble known as flat-pebble conglomerate. The rock fragments comprising them tend to be flat, current-aligned pebbles (Figure 4.6A). Early Ordovician flat-pebble conglomerates are widespread globally because large areas of submerged continental margins were concentrated in tropical latitudes where powerful hurricanes and storms were commonplace. Because the rock fragments within these deposits are generally the same lithology as the enclosing strata (limestone or dolomite), geologists sometimes describe the conglomerates as "intraformational." This means that the gravelly sediment was created, deposited, and lithified all in the same sedimentary environment.

FIGURE 4.6. (A) Flat-pebble conglomerate in the Fillmore Formation of the Ordovician Pogonip Group, Confusion Range, Utah. (B) Large-scale ripple marks and trace fossils preserved in the same formation.

There are several other indications within the Pogonip Group strata of both the strong currents and extremely shallow water covering the miogeocline. Large-scale ripple marks (Figure 4.6B) are commonly preserved on bedding planes of the various Pogonip Formations. These were formed when soft mud was heaved by oscillating tidal currents or directional bottom currents. During times of prolonged exposure of the miogeocline, fresh water from both land and tropical rains would seep into the soluble carbonate sediments. Limestone exposed at the surface was dissolved and washed away by rivulets of fresh water, and underground cavities formed in the subsurface as percolating water dissolved the calcite or dolomite in the carbonate strata. Eventually, these cavities and caverns were filled with rocky rubble when the voids collapsed. The dissolving of soluble rock at the surface and formation of underground cavities produces a unique set of geomorphic features known as karst topography, named for a portion of the Dinaric Alps between Italy and Slovenia where such features are well developed. Landforms typical of karst topography include solution valleys, sinkholes, and sinking rivers, while cavern systems characterize the underground realm in such regions.

Extensive karst features developed across the exposed Great Basin seafloor during several periods of prolonged sea level decline during the Ordovician Period. When the ancient karst surface was resubmerged by rising seas, these masses of cave rubble were lithified as irregular bodies and pockets of breccia in the limestone strata. Such masses of breccia have been observed in several different formations of the Pogonip Group and are thought to represent the lithified cave fillings of ancient cavern systems. Some of the bedding planes between strata of the Pogonip Group are also highly irregular with red-brown paleosols (ancient soils), providing additional evidence of karst development. Geologists use the term "paleokarst" (ancient karst topography features) to describe the cave-filling rubble and exposure surfaces preserved within the Pogonip Group.

Across the Great Basin, the strata of the Pogonip Group are overlain by one of the most distinctive rock units in the entire region: the Eureka Quartzite. Sandwiched between the dark-gray limestone and shale of the Pogonip Group below and the nearly black Late Ordovician dolomite sequences above, the light-tan Eureka Quartzite typically forms prominent cliffs and ledges that make it one of the most recognizable marker units in the Great Basin region (Figure 4.7). "Quartzite," in this formation, is a bit misleading because the term generally describes a metamorphic rock resulting from the alteration of sandstone. In fact, the Eureka Quartzite exhibits no sign of metamorphism.

FIGURE 4.7. Outcrops of the Eureka quartzite in the Confusion Range of western Utah. The hard Eureka quartzite generally forms light-colored cliffs beneath the nearly black Late Ordovician dolomite strata visible on the skyline.

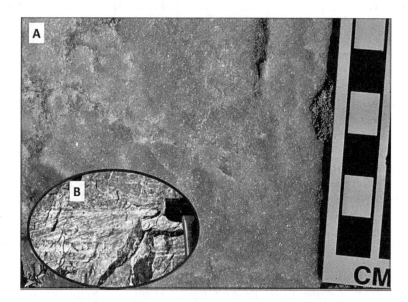

FIGURE 4.8. (A) Close-up of the fine-grained Eureka Quartzite from the Confusion Range of western Utah. (B, inset) Crossbedding in the Eureka Quartzite. Hammer head on the right is 12 cm (4 in) long.

Instead, it is an exceptionally pure, fine-grained sandstone that is well-lithified with quartz cement. In older sandstone classifications, the term "orthoquartzite" was used to describe such clean sandstones composed primarily of the mineral quartz. Unfortunately, geologists commonly drop the prefix "ortho" when describing these sandstones, letting the sedimentary context of the term's use distinguish it from metamorphic quartzite. The Eureka Quartzite is an excellent example of an orthoquartzite, as defined by earlier usage. While the modern term "quartz arenite" is less ambiguous for sandstone of this type, the Eureka Quartzite was named in 1883, long before the development of modern sandstone classification.

By any name, the Eureka Quartzite is a remarkable rock unit. It is the only formation in the thick Paleozoic succession of the Great Basin to consist entirely of granular sediment. It can be traced across 170,000 square miles of the Great Basin, with a typical thickness of about 250 feet and a maximum thickness of nearly 600 feet. The volume of sand contained in the Eureka Quartzite is a staggering 12,000 cubic miles. This fine sand is also exceptionally pure: quartz comprises more than 99 percent of the grains (Figure 4.8A). Only locally are noticeable amounts of other minerals present, typically feldspar and carbonates, but only as 1–2 percent of the grains in the rock. The sand grains are generally small (0.05–0.5 mm) and well rounded, indicating extensive transport from their source areas on land. Some of the grains in the Eureka quartzite are abraded and frosted, which suggests that wind played a role in their movement to the ancient Great Basin prior to deposition on the miogeocline. Preserved in the Eureka quartzite strata are several types of crossbedding, an indication of sand deposition under the influence of strong bottom currents (Figure 4.8b). The orientation of these crossbeds provides evidence that the sand moved into the Great Basin region from the modern north and northeast. This implies that the grains originated somewhere in eastern Laurentia and were transported by a combination of Ordovician trade winds and

the equatorial currents they sustained (Figure 4.9). Studies of the ages of zircon mineral grains in the Eureka Quartzite also suggest that the sediment was originally derived from ancient basement rocks in Canada, consistent with the paleocurrent reconstructions.

In most places in the Great Basin, the Eureka Quartzite rests unconformably on the carbonate rocks of the Pogonip Group. The basal Eureka unconformity reflects the period of exposure when the erosion occurred and karst features developed across the outer miogeocline. The post-Pogonip Group sea-level drop that uncovered the miogeocline may be related to the great regression terminating the Sauk sequence, or it may reflect other factors specific to the western edge of Laurentia. In any case, when the rising sea level resubmerged the Great Basin seafloor, a massive flood of sand from the still-exposed lands of Laurentia to the north and east spread across the Great Basin region. The Eureka Quartzite is the most prominent of several formations that document this surge of granular sediment into the shallow seas of the Great Basin. Other, approximately coeval, units that consist mostly of sand and silt include the Swan Peak Formation of central Utah and the Watson Ranch Quartzite to the west.

The sea-level rise that submerged the karst surface developed on the Pogonip Group and accompanied the flood of sand comprising the Eureka Quartzite signifies the transition from the Sauk sequence to the Tippecanoe sequence (Figure 4.2). From Middle Ordovician time through the Early Devonian Period, sea level rose erratically but ultimately submerged much of Laurentia. At the peak of the Tippecanoe inundation about 420 million years ago in the Late Silurian Period, the transgressing seas had drowned most of Laurentia. Only the highlands in the northeast corner of the ancient continent remained high and dry. Most of the Great Basin seafloor was once again submerged and would remain so for more than 50 million years during the Tippecanoe sequence. The Eureka quartzite was drowned. Carbonate deposition resumed and remained dominant from Late Ordovician to Middle Devonian time.

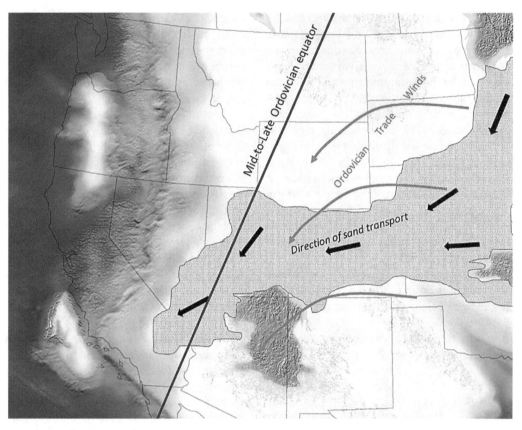

FIGURE 4.9. The Eureka Quartzite represents a flood of sand that swept across the miogeocline in mid-Ordovician time. The sand was transported from the northeast along the margin of flooded Laurentia by the Ordovician trade winds and equatorial currents.

But why did this great flood of pure sand wash across the miogeocline so abruptly in Middle Ordovician time? Geologists have identified several possible causes of this enigmatic event. The deposition of the Eureka Quartzite has been linked to intense volcanism in eastern Laurentia (and elsewhere) that may have resulted in acid rain downwind. The acidic precipitation would intensify weathering in central Laurentia, resulting in great volumes of granular quartz residue. This weathering debris was transported by wind and rivers to the western edge of the ancient continent. The rising sea level of the early Tippecanoe sequence stopped the westward transport of the quartz grains, and they piled up on the shallow Great Basin seafloor. Eventually, in later Ordovician time, the volcanic activity diminished, the acid rain

subsided, and carbonate sediments once again accumulated on the miogeocline. While there may have been other factors in the anomalous surge of sand to western Laurentia, it is probably more than a coincidence that it occurs at a time of distant volcanic outbursts coupled with rising sea level.

In the eastern Great Basin, the Eureka Quartzite is overlain by dark-colored dolomite strata generally assigned to both the Ely Springs and Fish Haven Dolomites. Westward, into central and southwestern Nevada, the dolomite sequences tend to grade into cherty limestone and shale in units such as the Hanson Creek and Mountain Springs Formations (Figures 4.4 and 4.10). The color contrast between the dark-gray to black dolomite and limestone sequences and the underlying light-tan Eureka

FIGURE 4.10. Outcrops of Late Ordovician strata in the Great Basin. Background photo illustrates the prominent color contrast between the black Ely Springs Dolomite and the underlying Eureka Quartzite in the Confusion Range of western Utah. Late Ordovician dolomite strata grade westward into limestone and chert such as that comprising the Hanson Creek Formation (*inset photo*) in the Monitor Range, central Nevada.

Quartzite makes the boundary between the two easy to spot wherever these rocks are exposed in Nevada and Utah. The dolomite and limestone formations signify a return to carbonate sedimentation across the miogeocline in Late Ordovician time, as sea level rose and the shoreline migrated east. The Late Ordovician rock sequences of the Great Basin are the oldest of the Tippecanoe sequence and accumulated in shallow, open marine settings. The dolomite that dominates these strata in the eastern part of the Great Basin is part of a persistent trend in the rock record of Middle Paleozoic time in the region. We will return to the origin and unique nature of these dolomite strata in the next chapter. However, it is important to note that the transition from dolomite in the eastern Great Basin to limestone and chert to the west is a pattern that persists well into the Silurian and

Devonian Periods and tells us something significant about the way massive dolomite deposits formed on the Great Basin seafloor.

Life and Death on the Ordovician Seafloor

A major increase in the diversity of life inhabiting the Great Basin seafloor occurred shortly after the beginning of Ordovician time. The survivors of the Cambrian extinction experienced a major pulse of evolution, giving rise to new types of trilobites, sponges, arthropods, brachiopods, and molluscs. These rapidly evolving Cambrian holdovers were joined by many new species of other invertebrate groups including graptolites, echinoderms, ostracods, corals, and primitive chordates. The result of this burst was an array of marine organisms profoundly different from anything that preceded it. So dramatic

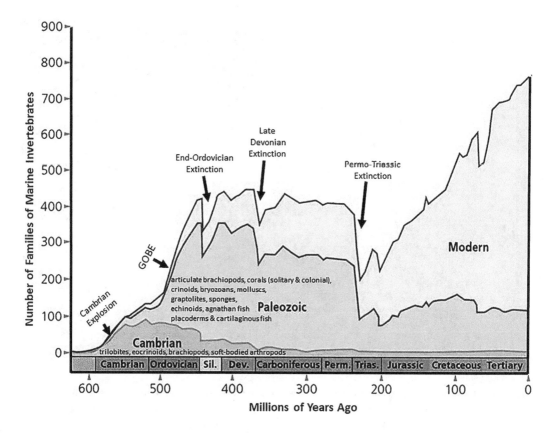

FIGURE 4.11. The Paleozoic fauna was established during the Great Ordovician Biodiversification Event (GOBE) and persisted for more than 200 million years until the Permo-Triassic extinction.

and widespread was the evolutionary riot on the Ordovician seafloor that scientists have given it a name: the Great Ordovician Biodiversification Event, or GOBE.

The GOBE was not instantaneous across the Great Basin seafloor. Rather, there appear to have been several waves of rapid evolution linked to the erratic environmental shifts that affected the Ordovician marine ecosystem. Not every group of marine invertebrates was affected at the same time or in the same manner as the GOBE played out over some 30 million years. However, when it finally culminated in the Late Ordovician, the multiple bursts of evolution had resulted in 4–5 times more genera (sing. genus) on the Great Basin seafloor than had existed following the "Cambrian explosion." The GOBE occurred at a time of rapidly changing global geography, rising levels of atmospheric oxy-

gen, fluctuating sea level, shifting climate, and intense tectonic activity. Any of these factors could have had significant influences on the shallow marine environment. In combination, their effects are still unknown, but it is probably more than a coincidence that such a great burst of evolution occurred against a backdrop of profound ecological shifts.

Environmental change is a powerful driver of evolution because it changes the rules of survival in natural habitats. Some groups of organisms fail to adapt to the new conditions and become extinct; others diversify into new lineages better suited to the new ecological conditions. During the GOBE, new and successful adaptations in the existing lineages of marine invertebrate arose much faster than the less-responsive groups died out. While the precise causes of the GOBE remain somewhat uncertain, they are

FIGURE 4.12. Preservation of Ordovician fossils from the Great Basin: (A) an example of rock from the Kanosh Shale composed mostly of well-preserved brachiopod shells and shell fragments; (B) a section of poorly preserved nautiloid shell from the Ely Springs Dolomite. Both examples are from the Confusion Range of western Utah.

almost certain to have been multiple, interconnected, and sufficient to incite one of the most dramatic surges of biodiversity in Earth history. At the conclusion, the overall structure of the marine ecosystem had been permanently modified from the relatively low-diversity Cambrian fauna to the dazzlingly diverse Paleozoic fauna (Figure 4.13). The major taxa comprising the marine invertebrate faunas of the world would then remain stable for the next 200 million years until the Permo-Triassic extinction. Near the end of the Ordovician Period, the GOBE was countered by one of the five greatest biotic crises to affect the oceanic realm—the End Ordovician extinctions (Figure 4.11). We will examine this extinction event in more detail later in this chapter. First, let's consider how the Ordovician fossil record of the Great Basin reflects proliferation of life inhabiting its seafloor some 460 million years ago.

Ordovician Fossils of the Great Basin

Fossils can be found in almost any exposure of the limestone and shale comprising the various formations in the Pogonip Group. In some places in western Utah (Kanosh Shale, Juab Limestone, Lehman Formation) and Nevada (Antelope Valley Formation), the abundance of Ordovician fossils borders on spectacular.

Such great concentrations of fossils likely originated when the storms and tidal surges that swept across the extremely shallow miogeocline washed heaps of shells into channels and low areas on the platform. Some horizons in the Pogonip strata are so richly fossiliferous that organic remains comprise the bulk of the rock (Figure 4.12A). In contrast, the clean sandstone in the Eureka Quartzite typically yields only trace fossils, mostly vertical cylindrical tubes left by burrowing invertebrates. With a few exceptions, the Late Ordovician dolomite formations produce fossils that are poorly preserved due to the secondary recrystallization of such carbonate rocks discussed in more detail in the following chapter (Figure 4.12B).

Many Ordovician fossils from Great Basin exposures are survivors of the Cambrian extinctions but appear as new species distinct from their ancestors. Trilobites are well represented in the Ordovician faunas of the Great Basin (Figure 4.13A, B) but are far less abundant and diverse than in the Cambrian. However, other groups of arthropods that were rare or nonexistent in the Cambrian explode during the GOBE. Of these, the ostracods, small bivalved crustaceans, are especially common in rocks of the Pogonip Group. Resembling little black or brown beans, ostracod fossils are so abundant

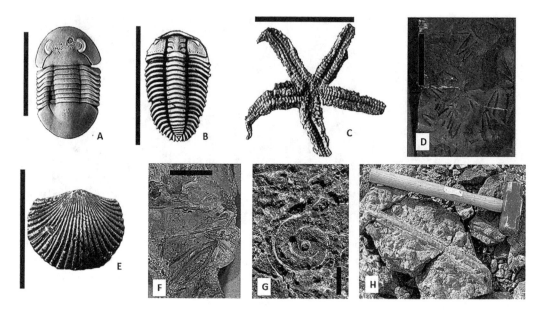

FIGURE 4.13. Representative Ordovician fossils from the Great Basin: (A) *Presbynileus*, a trilobite; (B) *Hintzeia*, a trilobite; (C) an asteroid echinoderm, or "sea star," from the Pogonip Group; (D) *Didymograptus*, a graptolite from the Swan Peak Formation; (E) *Shoshonorthis*, a brachiopod; (F) crinoid stalks from the Pogonip Group; (G) large gastropod (snail) from the Antelope Valley Limestone; (H) a section of the conical shell of a nautiloid from the Kanosh Shale. Scale bar = 3 cm in all, except H. Hammer handle in H is 40 cm/16 in long. A, B, C, and E courtesy of the Utah Geological Survey; all others from the author.

in some Ordovician strata of the Great Basin that their remains comprise over half of the bulk rock. More than 30 different species of ostracods have been identified from the Ordovician strata of western Utah alone. Among the echinoderms, the eocrinoids of the Cambrian Period are also present in Ordovician rocks, but these relatively primitive echinoderms are accompanied by a dazzling array of relatives including sea stars, crinoids, cystoids, and edrioasteroids.

Among the many new groups of organisms that flourished and diversified during the GOBE were the nautiloids, graptolites, bryozoans, and conodonts. The nautiloids were squid-like, tentacle-bearing molluscs that possessed simple conical shells, either straight or curved, with interior gas-filled chambers (Figures 4.13H, 4.14). Capable of regulating their buoyancy with internal gas pressure, the nautiloids propelled themselves through the seas by powerful water

jets expelled from the body. These molluscs possessed vicious plate-like teeth around the mouth and were predators that fed on a variety of sessile marine invertebrates.

Graptolites (Figures 4.13D, 4.14) were tiny suspension feeding animals that constructed colonies up to several inches in length. Each graptolite animal in the colony was housed in a separate tube-like pocket on a strand made of chitinous material. Graptolite fossils are normally preserved as carbon films along bedding planes that look somewhat like small hacksaw blades of varying size and shape. Most Paleozoic graptolite colonies were branched with one, two, four, or many stands. While some graptolites were sessile organisms, many lived as floating colonies, freely drifting in the near-surface water where their planktonic food was abundant. The remains of planktonic graptolites became widely dispersed and are preserved in

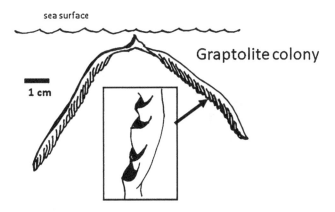

sea surface

Graptolite colony

1 cm

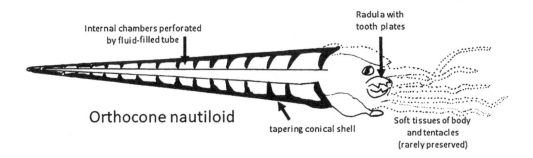

Internal chambers perforated
by fluid-filled tube

Radula with
tooth plates

Orthocone nautiloid

tapering conical shell

Soft tissues of body
and tentacles
(rarely preserved)

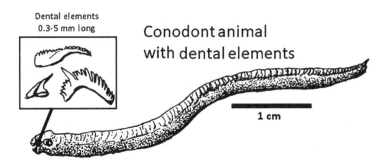

Dental elements
0.3-5 mm long

Conodont animal
with dental elements

1 cm

FIGURE 4.14. Some common nektonic (swimming) and planktonic (floating) animals of the Ordovician seas.

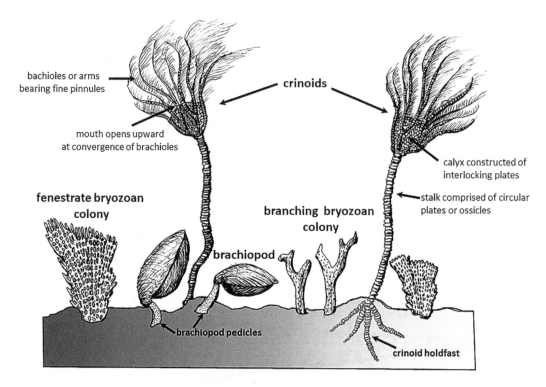

bachioles or arms
bearing fine pinnules

crinoids

mouth opens upward
at convergence of brachioles

calyx constructed of
interlocking plates

fenestrate bryozoan
colony

branching bryozoan
colony

stalk comprised of circular
plates or ossicles

brachiopod

brachiopod pedicles

crinoid holdfast

FIGURE 4.15. Some common Ordovician benthic invertebrates of the Great Basin seafloor.

fine-grained sediments across the Great Basin. This broad distribution makes them ideal for correlating Ordovician strata from distant localities. Graptolites were long thought to have died out in the Early Carboniferous Period, about 320 million years ago. However, the 2009 discovery near Bermuda of a new species belonging to the marine invertebrate class Pterobranchia raised some doubt about this assumption. This tiny (1/10 inch) animal builds colonies and possesses an anatomy so like that presumed for graptolites that it was named *Cephalodiscus graptolitoides* by its discoverer, Peter N. Dilly. Many scientists consider this pterobranch species to be a living graptolite, so perhaps the group of suspension feeders did not die out after all.

Conodonts (Figure 4.14) are tiny (generally no larger than a millimeter or two), tooth-like objects of many different shapes. For decades, scientists were uncertain what kind of animal left the conodont fossils, but the recent discovery of a dozen or so well-preserved specimens with body outlines have resolved that mystery

somewhat. It appears that conodonts belong to small, eel-like chordates that swam through the seas consuming prey such as floating invertebrate larvae and larger zooplankton. The conodont fossils are individual denticles made of calcium phosphate and comprised the complex mouth apparatus for these small predators. Denticles of various sizes and shapes were in different positions and arrangements in the mouth and throat of the conodont animal. After death and soft tissue decay, the dental elements fell out of the mouth apparatus and were commonly disassociated in seafloor sediments, which is why the identity of the conodont-bearing animals remained unknown for so long. These swimming predators are likely related in some way to modern jawless fish such as hagfish and lampreys. Conodont animals first evolved in the Cambrian Period, but their variety and complexity increased dramatically during the GOBE. Like the graptolites, conodont fossils are used extensively to correlate Ordovician rock sequences on both regional and global scales.

The graptolites, nautiloids, and conodonts mostly lived above the seafloor, either as nektonic (swimming) or planktonic (floating) consumers. On the current-swept shallow seafloor, a maze of other organisms filled a variety of ecological niches available on the miogeocline. The brachiopods and crinoids (Figure 4.13E, F) were sessile suspension feeders anchored to the seafloor by either a fleshy pedicle (brachiopods) or a root-like holdfast attached to a stalk made of circular plates (Figure 4.15). While most brachiopods appear superficially like clams (of the phylum Mollusca, class Bivalvia), they differ in that the two shells are not mirror images of each other. Rather, the plane of symmetry in brachiopods passes through the two opposing shells, not between them as in clams. The soft anatomy of these two groups of shellfish is likewise quite distinct. Brachiopods possess a spiral or loop-shaped organ called the lophophore that supports fine cilia and tentacles used to strain food from currents of water passing into the shells. Clams have no such organ to facilitate suspension feeding.

The crinoids are relatives of sea stars and sea urchins within the phylum Echinodermata. The body of the crinoid was a cup-shaped structure, the calyx, constructed from polygonal calcareous plates (Figure 4.15). The calyx was supported above the seafloor on a stalk consisting of a series of circular plates resembling a stack of tires. The mouth of crinoids was on the upper surface of the calyx, surrounded by arms (brachioles) supporting fine pinnules used to trap planktonic food. The organic matter collected by the array of brachioles was carried to the mouth by strands of mucus. The common name for modern crinoids, "sea lilies," is an apt description of the plant-like form of both living and fossil crinoids.

Within the phylum Echinodermata, another group to emerge during the GOBE were the sea stars, or "starfish," belonging to the class Asteroidea. While fossil sea stars are not abundant in the Ordovician strata of the Great Basin, some of the specimens from western Utah are spectacularly well-preserved (Figure 4.13C). Sea star exoskeletons are constructed from many calcareous plates that usually disaggregate after death, so complete fossils are rarely encountered in rocks of any age.

Yet another group of anchored suspension-feeders diversifying during Ordovician time were the bryozoans (phylum Bryozoa). Like living bryozoans, these small animals constructed calcium carbonate colonies but had much more complex soft anatomy than their coral mimics. Unlike corals, bryozoans possess a lophophore for feeding and respiration, a trait that relates them, at least distantly, to the brachiopods. Bryozoan colonies housed dozens to thousands of individual animals and were constructed in a variety of forms (Figure 4.15). Some bryozoan colonies were delicate networks with many open spaces that created a lacy appearance. More massive colonies were branching forms resembling stout twigs perforated by small openings. Some bryozoans also built encrusting colonies that grew directly on the seafloor or even on the shells and exoskeletons of other invertebrates such as crinoids or brachiopods.

Snails were also extremely abundant on the Great Basin seafloor in Ordovician time (Figure 4.13G). Snails are molluscs (phylum Mollusca, class Gastropoda) with a single coiled shell. Most Early Paleozoic snails possessed simple unadorned shells coiled in a nearly flat plane. The simplicity of most Ordovician gastropods is quite different from the modern marine snails, many of which secrete high-spired and elaborately ornamented shells. Early Paleozoic snails were mostly suspension feeders or grazing herbivores that fed on algal and/or cyanobacterial slimes on the seafloor. Ordovician snails were probably not active predators or deposit-feeders as are many modern marine snails. About a dozen different kinds of snails have been identified from Ordovician fossils in the Great Basin.

A variety of sponges (phylum Porifera) and taxonomically uncertain sponge-like organisms also inhabited the Great Basin seafloor during Ordovician time. Sponges generally form a hollow, basket-like body with perforated walls through which seawater circulates from the action of ciliated cells embedded in the sponge

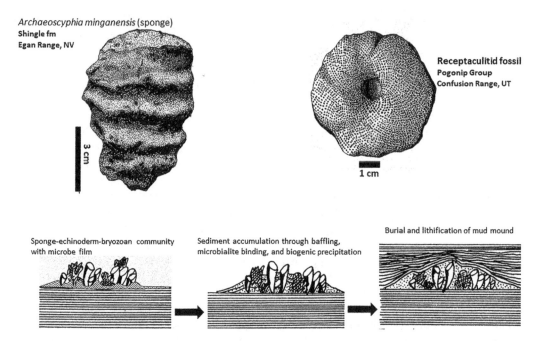

Archaeoscyphia minganensis (sponge)
Shingle fm
Egan Range, NV

3 cm

Receptaculitid fossil
Pogonip Group
Confusion Range, UT

1 cm

Burial and lithification of mud mound

Sponge-echinoderm-bryozoan community
with microbe film

Sediment accumulation through baffling,
microbialite binding, and biogenic precipitation

Origin of Ordovician Carbonate Mud Mounds

FIGURE 4.16. Ornamented Ordovician sponge (*top left*) and the typical form of a fossil receptaculid (*top right*) from the Great Basin. These sessile organisms are commonly associated with reef-like mounds preserved in the Ordovician strata of the Great Basin seafloor.

tissues. Most sponges utilize tiny spikes, multi-armed spicules, rods, or plates of silica or carbonate minerals to support the soft tissues of the walls and keep it raised above the seafloor. These skeletal components, usually scattered from the sponge body, are common as fossils in the Ordovician strata across the Great Basin. In many exposures of fossiliferous Ordovician strata, the entire body of sponges, or large portions of it, are preserved. Like their modern descendants, Ordovician sponges had a variety of shapes. Many were shaped like simple conical baskets, with the larger opening directed upward from the seafloor. Others had a more flattened, bulging, or biscuit-like, shape, and still others had wrinkles and ridges on the outer walls (Figure 4.16).

Among the widespread sponge-like organisms was a group known as the receptaculitids, named for the common Devonian genus *Receptaculites*. Receptaculitids have such a unique skeletal structure that paleontologists are still

not entirely sure how to classify them. Some scientists consider them early sponges, while others have suggested a relationship with corals or even calcareous algae. A few paleontologists have suggested that the receptaculitids be assigned their own taxonomic category, phylum Receptaculita. This uncertainty reflects the very distinctive symmetry and architecture of the calcified exoskeleton. Whatever they were, the receptaculitids secreted a double-walled, generally bowl- or tube-shaped mineralized carapace. The outer surface of the perforated structure was a mosaic of small plates arranged in intersecting spirals radiating outward from the center, somewhat like the seeds in a sunflower (Figure 4.16). This array of plates and intervening pores gave the outer surface of the receptaculitid exoskeleton an appearance like a pin cushion. Because no major group of living marine invertebrates has a similar radial spiral structure, no modern analogues are available for comparison with the Ordovician fossils. Yet, although the feeding

FIGURE 4.17. Large carbonate mud mound (dashed line) in Ordovician limestone at Meiklejohn Peak, Nevada.

mode and internal anatomy of the receptaculitids remain unknown, most paleontologists surmise that they were suspension feeders, as were the bryozoans and crinoids with which their remains are commonly preserved.

Another enigmatic fossil common in the Great Basin Ordovician strata is *Calathium*, a genus with several species of organisms that secreted a perforated, double-walled exoskeleton like the sponges and receptaculitids. The overall structure of the *Calathium* exoskeleton varied from basket-shaped to tubular to vase-shaped. It was anchored to the seafloor by a primitive holdfast and appears to have opened upward into the water column. These characteristics suggest that *Calathium* was a primitive sponge, but some scientists have concluded that it may be a support structure secreted by calcareous algae or a receptaculitid.

In the Pogonip Group and correlative strata of the Great Basin, fossils of sponges and sponge-like organisms commonly occur in dense concentrations associated with other suspension-feeding invertebrates, microbial laminations, and stromatolites. This suggests

that the sponges, along with crinoids, bryozoans, and echinoderms, tended to colonize seafloor surfaces stabilized by cyanobacterial growth. Attached to the seafloor coated by bacterial slime, these "thickets" of suspension feeders could resist, to some degree, the strong currents that swept across the Ordovician miogeocline. In many areas across the shallow miogeocline of eastern Nevada and western Utah, the clusters of suspension-feeding organisms grew into small, reef-like masses or "patch reefs." As they developed, the patch reefs acted as sediment baffles, trapping silt and mud carried across the seafloor by strong bottom currents. The mud accumulated in a mound that surrounded, and eventually buried, the organic framework, resulting in the formation of carbonate mud mounds preserved in many exposures of Ordovician strata in the Great Basin (Figures 4.16, 4.17). Many of these mud mounds are elongated roughly parallel to the modern northeast–southwest trending shoreline in central Utah. This alignment suggests that persistent longshore currents, driven by the Ordovician trade winds, continuously swept across the miogeocline, shaping the mud

FIGURE 4.18. Close-up of Ordovician limestone of the Antelope Springs Formation, Arrow Canyon Range, Nevada. Note the irregular masses of white calcite filling voids dissolved by rainwater during low tides. Pen is about 15 cm long.

FIGURE 4.19. Reconstruction of the Paleozoic fauna of the Great Basin in Ordovician time. Grapto-lite colonies float in the extremely shallow water above the Great Basin seafloor carpeted with brachiopods, crinoids (*background*), twig-shaped bryozoans (*middle left*), sponges (*middle right*), trilobites, gastropods (*lower left*), microbial mats, and receptaculitids (*lower right*). A nautiloid cephalopod (center, ~40 cm/16 in long) preys on these benthic organisms. In the distance, mud mounds and islands rise from the shallow seafloor. Original painting by Carel Brest van Kempen.

mounds and patch reefs into elongated parallel hillocks. Some of the larger mounds that developed in especially shallow water were exposed periodically during the lowest tides or when sea level fell slightly. As a result, the limestone in the mud mounds commonly has mineral-filled cavities that originated as open pockets formed through dissolution of carbonate minerals by rainwater during times of exposure. When the mud mounds were resubmerged, white calcite was deposited in the open pockets to create the irregular white blotches laced through the gray limestone (Figure 4.18).

Compared to the relatively low-diversity Cambrian fauna that preceded it, the Paleozoic fauna established during the GOBE was dazzlingly diverse (Figure 4.19). The surge of rapid evolution that gave rise to such variety of life in the Ordovician Period was a global phenomenon and is well documented by the fossils of this age in the Great Basin. However, near the end of the Ordovician Period, the GOBE was countered by one of the five greatest biotic crises to affect the oceanic realm: the end-Ordovician extinctions (Figure 4.11). The abrupt stop of the GOBE, and the dramatic loss of marine biodiversity during the wave of extinction that followed it, is also well recorded. Though there is still some mystery surrounding the end-Ordovician extinctions, the Great Basin strata and fossils have provided some clues to what events could have triggered such a biotic calamity on the seafloor.

Ordovician Extinction and Ice Ages

Rocks that formed in the final 1.5 million years of Ordovician time, between 445.2 to about 443.8 million years ago, comprise the Hirnantian Stage, the last of several subdivisions of the rock record for that period. It was a time marked by two significant global events, both of which are recorded in the strata of the Great Basin: a major mass extinction and a severe ice age. The Hirnantian ice age appears to have resulted from several interrelated events and circumstances that collectively plunged the earth into a short but bitterly cold glaciation. First, at the onset of the ice age, much of the land area of

the earth was centered on the South Pole as the ancient supercontinent of Gondwana (Figure 4.1). Gondwana was massive, consisting of what is today southern Europe, Africa, South America, Antarctica, and Australia. Laurentia was one of only a few major landmasses that were not a part of Gondwana in the Late Ordovician. Gondwana had been drifting toward the south pole throughout the Ordovician Period, but by Hirnantian time enough land had arrived in this position to provide a surface on which snow could collect rather than falling into open sea. As snow blanketed the expansive Gondwana landscape, more of the feeble energy received from the sun was reflected to space. The temperature fell, more snow accumulated, less solar energy was absorbed by the highly reflective snow, and the temperature plummeted even further. This is a classic example of a positive feedback loop in a natural system: one initial event (cooling related to Gondwana arriving on the south pole) triggers others (more snow, more reflected energy), which accelerates the original shift. The movement of Gondwana was not the sole cause of the Hirnantian glaciation, but it set the stage for the rapid deterioration of the global climate that would soon follow.

In addition to Gondwana's gradual drift south in the Early Paleozoic Era and its steady growth through the convergence of numerous plates at its margins, the tectonic crashes resulted in volcanism, deformation, and mountain-building along the suture zones. Oceanic rocks caught up in those zones came to the surface as the contorted and crumpled cores of rising mountain ranges. Prior to and during the collisions, volcanoes emitted lava that cooled to form new rock adjacent to the rising mountains. Once they were created or lifted in the mountainous regions, the newly exposed rocks began to weather. The weathering of rocks consumes atmospheric CO_2 and reduces greenhouse warming. Geologists have estimated that the enhanced weathering related to Ordovician mountain-building in Gondwana and elsewhere may have reduced levels of atmospheric CO_2 by nearly 50 percent. This reduction in the primary greenhouse gas in the atmosphere caused

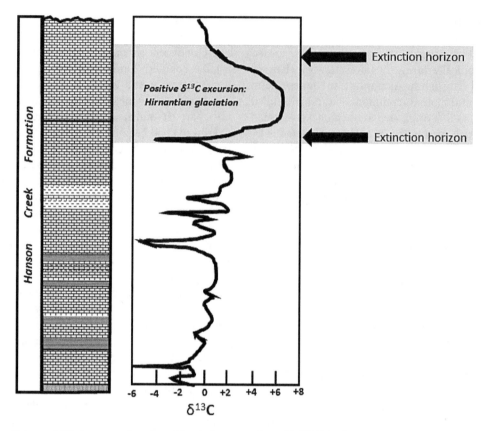

FIGURE 4.20. Increased carbon-13 relative to carbon-12 ($\delta^{13}C$) in the Late Ordovician Hanson Creek Formation of the Monitor Range, Nevada, is an indication of the Hirnantian glaciation. Note the correlation between this geochemical anomaly and the pulses of extinction. $\delta^{13}C$ data from Jones et al. (2017).

the global climate to cool even faster. Finally, as ice expanded on Gondwana, water was transferred from the oceans to the growing ice sheets, and sea level fell. The eustatic drop in sea level exposed more rock to weathering around the periphery of all continents, further increasing the drawdown of atmospheric CO_2. The earth got colder, and colder, and colder. During the last few million years of the Ordovician Period, the Hirnantian ice age was in full bloom.

The Late Ordovician rocks of the Great Basin record the Hirnantian glaciation in several ways. The drop in sea level caused by the expansion of ice on land resulted in a major regression around the margins of Laurentia. As the seas covering the Great Basin seafloor withdrew to the west, Late Ordovician sediments in the Fish

Haven and Ely Springs Dolomite of western and central Utah were exposed to erosion. This exposure in Hirnantian time resulted in a widespread erosional unconformity that separates these Late Ordovician sediments from overlying Silurian strata. Farther west, in the deeper portion of the miogeocline that remained submerged, sandy sediments were dispersed from the continental interior over tracts of the Great Basin seafloor where carbonate rocks had been deposited earlier. Such sandy intervals marking the Hirnantian regression have been identified in the Hanson Creek Formation and coeval rock units in eastern and central Nevada. Still farther west, where siliceous shale and chert has accumulated in deeper-water basins, fine-grained gray limestones appear, as water depth was reduced by the falling sea level. This shift to shallow-water carbonate sediments is recorded in the Vinini Formation of central Nevada.

In addition to these lithologic trends, geochemical anomalies have also been discovered in the latest Ordovician strata of the Great Basin that further reflect consequences of the Hirnantian ice age. An enrichment in Carbon-13 over normal Carbon-12 (technically, a positive $\delta^{13}C$ excursion) in carbonate rocks has been detected in the several different Late Ordovician formations in Nevada and western Utah (Figure 4.20). This isotopic shift is most likely attributable to falling sea temperatures, coupled with lower organic production in the seas and rapid burial of organic carbon, as planktonic organisms that were adapted to warm waters begin to die off. Geochemical analyses of Late Ordovician rocks in the Monitor Range of Nevada have also uncovered an intriguing increase in the concentration of mercury, a toxic metal, just as the Hirnantian glaciation began. Mercury levels spiked as much as 500 times the normal level in the water covering the Great Basin seafloor, then quickly returned to normal levels. Geologists interpret this spike as a reflection of a volcanic outburst, either on land or along the midocean ridges. Wherever the mercury may have originated, its sudden spike added even more ecological stress to the organisms populating the Late Ordovician seas off western Laurentia. The environmental volatility resulting from the Hirnantian glaciation and coeval ecological disturbances triggered several waves of extermination that ultimately reshaped life on the Great Basin seafloor: the end-Ordovician extinction.

The end-Ordovician extinction was not a sudden, singular die-off of marine organisms. Instead, there were several waves of extinction over the final 5 million years of the Ordovician Period. But their cumulative effect was a stunning crash in the biodiversity that arose during the GOBE (Figure 4.11). Paleontologists estimate that, overall, about 80 percent of all species, 50–60 percent of all genera, and 25 percent of all families of invertebrates vanished during this great extinction. Not all groups of organisms living on the Great Basin seafloor were affected at the same time, or to the same degree, by the end-Ordovician extinction. Trilobites, brachiopods, conodonts, and graptolites were the principal victims of the great die-off. About two thirds of all trilobite families disappeared, and this group of arthropods never really recovered. Though a few groups of trilobites survived to the end of the Paleozoic Era, they never again dominated marine faunas after the Ordovician Period. Some 70 percent of the conodont genera vanished during the end-Ordovician extinction, while about a third of all brachiopod families died out. But there were also some groups of organisms in the Paleozoic fauna that, for some reason, were less susceptible to extinction. The bryozoans lost only about a tenth of the species present during the earlier Ordovician, while the crinoids and cephalopods were also less drastically affected.

The precise causes of the end-Ordovician extinctions remain unknown, but it is likely that several factors may have played roles in initiating this great collapse of marine biodiversity. First, vast areas of the continental shelves surrounding Laurentia and other continents were exposed as sea level fell. In the Great Basin, this means that the shallow miogeocline habitat to which most Ordovician sea life was adapted was strongly reduced. As in the modern world, the loss of critical habitat intensified competition and, ultimately, led to the extinction of some lineages. Also, sea surface temperatures fell during the Hirnantian glaciation by as much as 10°C, adding more stress to mostly warm-water Ordovician creatures. The decline in surface temperatures would have strongly affected planktonic organisms, including the photosynthetic phytoplankton that were the foundation of the marine food web. As these primary producers died off in great numbers, less food was available to the numerous groups of Ordovician suspension-feeders living on the seafloor.

In addition, the decomposition of residual organic matter settling to the deeper seafloor consumed oxygen from bottom waters and from the soft sediment. Some scientists have further suggested that undersea volcanism may have warmed the seawater locally, driving out even more oxygen and expanding the deep-water anoxic zone. As local and ephemeral anoxic conditions developed in the deep ocean environment from multiple causes, the survival of benthic organisms became increasingly difficult.

Finally, just as the surviving organisms were beginning to adapt to the new conditions, the short-lived Hirnantian glaciation abruptly ended, sea level rose as terrestrial ice melted, and the surface water warmed again. This rapid environmental shift is no doubt related to the later phases of the end-Ordovician extinction that seem to disproportionately affect cool-water species.

What is certain about the end-Ordovician extinction is that it was a biotic response to a wild ecological roller coaster caused by a combination of factors unique in Earth history. There were other mass extinctions, one even more severe, but none appear to be linked to the same combination of factors that led to the die-off at the end of the Ordovician. When the Hirnantian glaciation ended, and rising sea level submerged the Great Basin seafloor, sediment again accumulated in stable, shallow marine environments.

However, the pattern of deposition after the Late Ordovician is so different from anything preceding that it poses an enduring geological mystery: the infamous "dolomite problem," a conundrum that has confounded Great Basin geologists for decades.

5

The Dolomite Interval

A Carbonate Conundrum

Once the Hirnantian ice age ended and rising seas again submerged the Great Basin seafloor, the world entered a long period of relatively warm but still somewhat variable climate that started near the beginning of the Silurian Period (444–419 million years ago). On land, the massive glaciers generally retreated under the warming conditions but did so erratically, as the warm intervals were punctuated by brief cold interludes. Nonetheless, the overall warming trend culminated in a world that was nearly ice free by Middle Silurian time. Global sea level rose as glacial meltwater drained into the oceans, and shallow seas transgressed continental margins and shelves worldwide.

The transgression that flooded most of Laurentia marked the beginning of the Tippecanoe sequence of rocks that can be recognized throughout North America. Most of the world's land area was in the southern hemisphere, where the gigantic supercontinent of Gondwana still stretched from the equator to the south pole. The vast Panthalassa super ocean covered most of the northern hemisphere, with three smaller continents positioned near the equator (Figure 5.1A). Laurentia was one of those equatorial continents, with the future Great Basin region situated near the northwestern margin, a little south of the equator, facing the Panthalassa Ocean. The eastern side of Laurentia was an eroded, still mountainous highland bordered by a narrow seaway, the Iapetus Sea. The

seaway narrowed throughout Silurian time, as the three equatorial continents converged. Ultimately, this convergence would lead to a tectonic collision in Devonian time that initiated a major period of mountain building along what is now the northern Appalachian Mountains of North America.

During most of the Silurian Period, the shoreline was in central Utah, and the Great Basin seafloor sloped gradually toward the modern west (north, according to Silurian geography) and the deeper eugeocline basin situated in western Nevada and eastern California (Figure 5.1B). The edge of the shallow miogeocline ran through central Nevada in roughly the same position and following the same trend as in the Ordovician Period. In other words, the conditions across the Great Basin seafloor through Silurian time and into the early part of the Devonian Period were not dramatically different from what was established during the Early Paleozoic era.

However, despite these similarities in paleogeography and environmental conditions, the Late Ordovician to Early Devonian rock record of the Great Basin is quite different from strata that formed earlier or later. How did the pattern of sedimentation change during this interval? The answer introduces one of the most stubborn mysteries in Great Basin geology, a puzzle that geologists have still not entirely solved after more than a century of study.

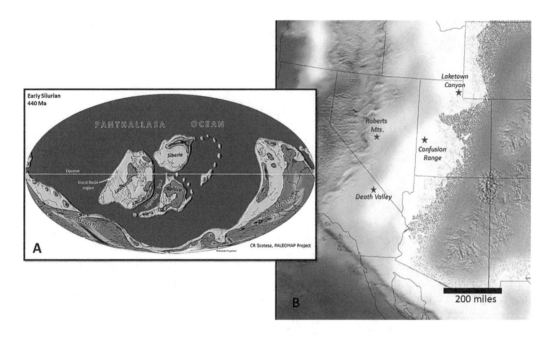

FIGURE 5.1. Global geography in the Early Silurian Period (A) and Silurian geography of the Great Basin region about 430 Ma (B). A: base map from Scotese and Wright (2018), PALEOMAP Project (https://www.earthbyte.org/paleodem-resource); B: paleogeographic base map prepared by R. Blakey, Colorado Plateau Geosystems Inc.

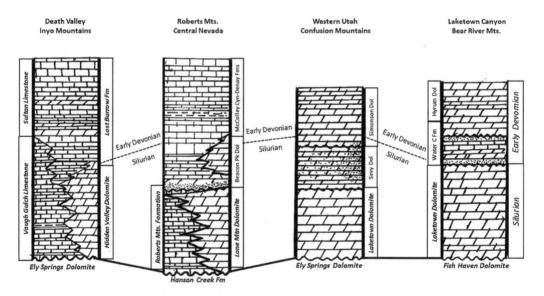

FIGURE 5.2. Generalized Silurian and Early Devonian rocks across the Great Basin. The columns match the locations depicted in Figure 5.1B.

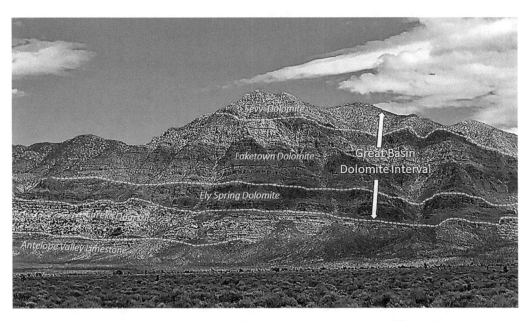

FIGURE 5.3. Formations of the Great Basin dolomite interval exposed on the west slope of the Mount Irish Range in southern Nevada.

Silurian-Early Devonian Rocks of the Great Basin: The Dolomite Dilemma

We learned in the preceding chapter that the carbonate rock dolomite (or dolostone, as some geologists prefer) became dominant in the Late Ordovician Period across the eastern Great Basin. This trend not only continues into Silurian and Early Devonian time but is enhanced to the point that dolomite comprises more than 90 percent of the regional rock record. In formations such as the Laketown Dolomite, the Sevy Dolomite, and the Simonson Dolomite of central and western Utah, dolomite strata are collectively more than 3,000 feet thick (Figure 5.3). As their names indicate, these thick sediments consist almost entirely of dolomite, with minor amounts of silty limestone. The stratified rocks in each unit tend to be thick-bedded to massive, and commonly weather to form bold brownish-gray to orange-tan cliffs. In central Utah, where deposition occurred very near the coastline, the deposits are relatively impure, containing sandy and silty material interspersed with the dolomite. Formations similar in composition to those first named in Utah can be recognized across a broad region stretching from Idaho to southern Nevada and can be correlated with rock sequences as far west as central Nevada. In the Roberts Mountains region of Eureka County, for example, Silurian–Early Devonian dolomite strata compose most of the Lone Mountain Dolomite and Beacon Peak Dolomite (Figure 5.2). Overall, the volume of dolomite in these widespread rock units is staggering, especially if we consider the underlying Late Ordovician dolomite as part of the sequence. Unlike the more heterogenous rock record of earlier Paleozoic time, a single sedimentary lithology dominates the successions in the eastern Great Basin from Late Ordovician to Early Devonian time—a span of more than 55 million years. This period, from about 450 to 395 Ma, is the Great Basin dolomite interval. In no other portion of the immense geologic history of the region is dolomite, or any other single rock type, so dominant.

Though dolomite overshadows all other sediment types in Silurian–Early Devonian exposures in the eastern Great Basin, these

FIGURE 5.4. Nearly vertical beds of Silurian limestone in the Vaughn Gulch Formation of the Inyo Mountains, eastern California, are typical of rocks that formed in the western Great Basin during the dolomite interval.

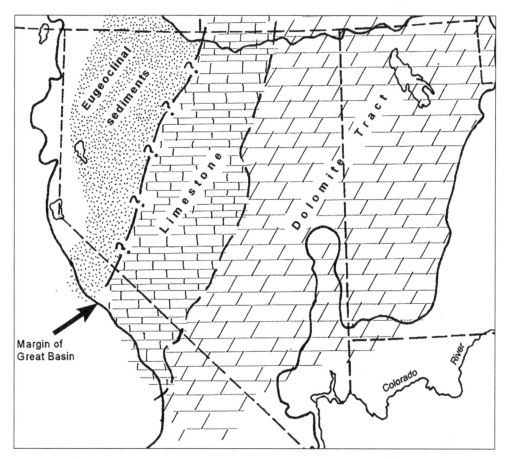

FIGURE 5.5. Distribution of dolomite and limestone across the Great Basin seafloor during the dolomite interval, Late Ordovician to Early Devonian time. Position of the eugeocline in the far west is uncertain due to eastward displacement of the strata along later Paleozoic thrust faults.

strata grade into limestones and chert to the west. From central Nevada westward, limestone becomes the dominant lithology in rock units such as the Roberts Mountains and Vaughn Gulch Formations (Figure 5.4). Subordinate layers or nodules of chert, siltstone, and fine sandstone commonly occur in the limestone-dominated formations. The transition from nearly pure dolomite in the east to more heterogenous, limestone-dominated sequences to the west is so persistent that the miogeocline can be divided into a near-shore dolomite tract and an offshore limestone tract (Figure 5.5). This distribution suggests that dolomite sediments generally formed in the shallowest nearshore environments, while the limestone sediments characterized the more offshore, open-water settings on the ramp-like miogeocline. Beyond the edge of the continental shelf, a deep-water eugeoclinal basin must have existed in Silurian and Early Devonian time. However, the precise location of this basin is uncertain because the deep-water sediments that formed there were later displaced eastward over the shelf along thrust faults accompanying Middle Paleozoic tectonic events.

Recall from Chapter 1 that dolomite and limestone are both carbonate sedimentary rocks that form in similar environments. The difference between these two rocks is their mineralogy: limestone consists mostly of calcite ($CaCO_3$), while dolomite is a double carbonate of calcium and magnesium, $MgCa(CO_3)_2$. It might seem a simple matter to add a little magnesium to calcite to form dolomite. However, the precipitation of dolomite from normal seawater is far from simple and appears to be limited to unusual circumstances and special chemical and physical conditions. Thus, the origin of massive and widespread dolomite deposits such as those in the Great Basin is an uncertainty that has confounded generations of geologists. Yet, in our attempt to unravel the deep history of the Great Basin seafloor, the significance of the Late Ordovician to Early Devonian dolomite is critical. What does this rock tell us about the conditions of the time? How are we to interpret the overwhelming dominance and singularity of this very distinctive rock, and the time interval during which it formed? To answer these questions, we need to consider in a bit more detail the mystery of dolomite formation. Why, after all, is the "dolomite problem" so problematic?

The Dolomite Problem in the Great Basin

Simply stated, the dolomite problem is a conundrum for geologists because nowhere in the modern world do massive dolomite deposits form. Only in very unusual natural environments such as hypersaline tidal flats in hot climates, warm salt lagoons, or where carbonate rocks have been affected by hydrothermal solutions, is dolomite produced. Even then, only small amounts, not massive and extensive deposits, form from direct precipitation. The mystery is only deepened by laboratory experiments demonstrating that, while the primary precipitation of dolomite from a solution is possible, it only happens under very specific and abnormal chemical conditions. In such experiments, small crystals of dolomite have been produced from very alkaline solutions of just the right chemistry and at high temperatures. Nor did these conditions exist across the expansive dolomite tract on the Great Basin seafloor during Late Ordovician to Early Devonian time. Then how did such enormous volumes of this mineral accumulate on the seafloor during the Great Basin dolomite interval? To formulate a plausible explanation, we need to consider the chemistry of carbonate minerals in just a bit more detail.

The carbonate chemistry of the oceans is complex, but one of the most common ways to form solid crystals from seawater is through the following reaction:

$$2HCO_3^- + Ca^{+2} \rightleftharpoons CaCO_3 + H_2CO_3$$
(ions in solution)   (calcite) (carbonic acid)

Note that this reaction is reversible; any addition of components or ions (the electrically charged atoms or group of atoms on the left) to one side of the equation drives it in the opposite direction. For sake of simplicity, we will omit consideration of the sources of the ions and

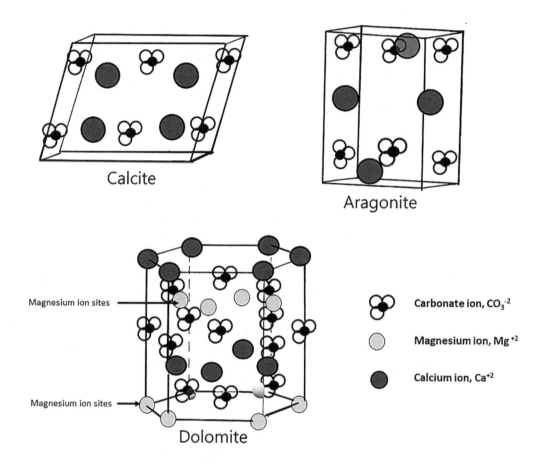

Magnesium ion sites

Magnesium ion sites

Carbonate ion, CO_3^{-2}

Magnesium ion, Mg^{+2}

Calcium ion, Ca^{+2}

FIGURE 5.6. Simplified atomic structure of calcite, aragonite, and dolomite.

factors that control the concentration of the reagents (they are multiple, complex, and vary from place to place and over time) but simply observe that calcium carbonate is continually crystallizing from, and dissolving in, seawater.

When calcium carbonate mineral crystals precipitate from seawater, the two most common forms are calcite or aragonite. The two minerals have the same composition, $CaCO_3$, but differ in the arrangement of ions and thus a different crystal symmetry: calcite is rhombohedral, while aragonite is orthorhombic (Figure 5.6). Several factors determine which is dominant in carbonate sediments and rocks. Remember also that organic carbonate secreted as skeletal or shell material by organisms is an important source of sediment in limestone. Some marine invertebrates, such as calcareous algae, some molluscs, and modern corals, secrete aragonite,

while others (gastropods, bryozoans, brachiopods, etc.) build skeletal material from calcite. The rate of crystal growth can also influence which mineral is formed, with rapid crystallization from solution favoring aragonite. Finally, time is a factor because calcite is more stable in the geological environment than aragonite. Over time, aragonite tends to be converted to calcite as the atoms and ions rearrange themselves into a more thermodynamically stable and permanent configuration.

To make dolomite, magnesium ions must be incorporated into the atomic lattice along with the calcium that is present in calcite or aragonite. This might seem possible by simple substitution of ions for Ca^{+2} ions. In nature, however, this is not so simple. While the two ions have the same electrical charge, they differ significantly in size. The ionic radius of the Ca^{+2} ion is about

FIGURE 5.7. Magnified view of crystalline dolomite from the Silurian Laketown Dolomite of western Utah.

114 pm (pm = picometer = 10^{-12} or 1 trillionth of a meter), while the radius for the Mg^{+2} ion is 86 pm, about 26 percent smaller. This means that the magnesium ion does not fit perfectly into the location in the atomic lattice occupied by calcium ions in aragonite and dolomite. The difference in size requires a rearrangement of the atomic lattice as depicted in Figure 5.6. Moreover, the concentration of the two ions is highly variable from place to place in the seas. Magnesium is typically more abundant than calcium in seawater because the latter, used extensively by organisms, cycles between the hydrosphere, biosphere, and lithosphere more readily. Depending on the marine biota and physical environment, calcium is usually more prevalent in carbonate minerals than magnesium. These limiting factors notwithstanding, there can be some minor substitution of magnesium for calcium in certain depositional settings or by organisms, resulting in "high-magnesian" calcite. Because the magnesium ions are few and randomly placed in the atomic lattice, this is not dolomite. Dolomite, a double carbonate, contains Mg^{+2} ions in specific zones or layers within the atomic lattice alternating with other layers of Ca^{+2} ions (Figure 5.6).

Precisely how the highly ordered structure of dolomite develops is the main uncertainty underlying the "problem." Again, there is nowhere in the modern world with massive amounts of well-ordered dolomite. One can find very minor quantities, usually small nodules or a thin crust, in restricted marine environments such as hypersaline tidal flats and inlets. Geologists have also discovered limited amounts of a poorly ordered variant called "proto-dolomite" in some marginal marine settings. Yet in the Great Basin, there are literal mountains of this mineral in Late Ordovician and Silurian strata.

A specific and complete solution to the dolomite problem, one that satisfies all geologists, is still being pursued and debated. However, studies of the Ordovician to Early Devonian dolomite on the Great Basin seafloor are providing important clues to resolving the long-standing mystery. First, the Great Basin strata are commonly composed of relatively large intergrown crystals of dolomite (Figure 5.7). Laboratory experiments have demonstrated that such crystals are unlikely to have formed by direct precipitation from seawater. Instead, they appear to have been the product of very slow crystallization. Many dolomite samples also exhibit tiny pockets and cavities, suggesting that fluids migrating through the sediment may have played a role in the crystallization process. Furthermore, poorly preserved relicts of brachiopod and coral shells, originally secreted as calcite or aragonite but now comprised of crystalline dolomite, have been observed in the Great Basin dolomites. This suggests that calcite can be converted, or recrystallized, into dolomite over long periods. Even the inorganic carbonate crystals in Great Basin dolomite sometimes exhibit vestiges of the calcite structure, although they now consist of dolomite.

These observations support the hypothesis that massive dolomite is not a primary deposit but was very slowly converted from calcite or aragonite. The recrystallization process may involve ion-by-ion replacement of calcium with magnesium, along with a corresponding rearrangement of the atomic lattice. Could this be the reason that dolomite is so rare in the modern world but so prevalent in parts of the ancient rock record? If so, then the pivotal question is how the conversion process, known as "dolomitization," occurs and why it was so pervasive during the Great Basin dolomite interval. Once again, the strata of the Great Basin seafloor offer a clue.

We have already seen that the massive dolomite accumulations in the Great Basin are restricted to the inner portion of the miogeocline, situated in what is now eastern Nevada and western Utah (Figure 5.5). Because the Silurian shoreline ran through central Utah, this distribution implies that the process of dolomitization was mostly a near-shore phenomenon that did not affect the carbonate rocks accumulating on the outer portion of the miogeocline. There are several possible explanations for this pattern, some of which can be evaluated by studying marginal marine settings in the modern world where the small amounts of dolomite have been found forming in near-shore environments. One such place is the Middle East and Mediterranean region, where low coastal basins are submerged by seawater during the highest tides (Figure 5.8). These low-lying intertidal or supratidal basins are known as sabkhas

(also spelled sebkha), a general Arabic term for "salt flat." In these areas, the normal seawater is frequently isolated from the open ocean at low tide, and strong evaporation in the arid climate creates dense hypersaline brines in the supratidal basins. Salt is commonly deposited on the sabkha surface as the salinity of the trapped seawater rises during evaporation. As the salt crystallizes, the residual briny water contains high concentrations of many substances, including Mg^{+2} ions. Their elevated density results in the downward and offshore seepage of the reactive fluid through previously deposited sediments (Figure 5.8). Over time, the movement of magnesium-rich fluids through the carbonate mud can gradually foster the replacement of calcium ions with magnesium ions at specific sites in the atomic structure of calcite. In this concept, known as the evaporative reflux model, the slow formation of dolomite requires the flow of hypersaline fluids through carbonate deposits over long periods of time. This may help explain why only very small amounts of dolomite are found in modern sabkha environments.

However, if the reflux were sustained over long periods of geological time, it is conceivable that great volumes of calcite sediment could be converted to dolomite in tracts of the seafloor closest to the coast. Dolomitization could even take place after lithification of the carbonate muds into limestone, provided that the fluids continue to migrate through the rock. Because the dolomitizing fluids are most effective nearest the shore, the dolomite tract would likely pass offshore into normal carbonate mud in which calcite remains unaltered. This pattern precisely matches the distribution trends observed in the strata of the Great Basin dolomite interval.

The evaporative reflux model for the origin of Great Basin dolomite is a plausible explanation, but not a perfect one. In the modern Middle East and Mediterranean regions, great volumes of salt (geologically known as the mineral halite) coat the surface of the sabkha basins as the hypersaline brines evaporate or seep into the subsurface. These salt deposits should have accumulated in, and landward of, the dolomite tract on the Great Basin seafloor. There are no

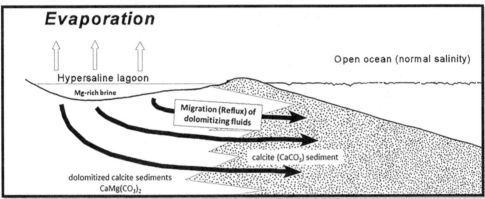

FIGURE 5.8. *Upper*: view from space of Sabkha El Melah in Tunisia. The salt-encrusted basin enclosed by the dashed line is periodically submerged during the highest tides. The sabhka is about 13 miles long from west (*left*) to east (*right*); *lower*: evaporative reflux model for the formation of dolomite in carbonate mud deposits. Upper image from NASA Landsat 7 image gallery.

extensive salt deposits in the Late Ordovician to Early Devonian strata of central Utah or regions to the east, but the reflux model may still be valid because the oceans did not continuously submerge western Laurentia beyond central Utah during the dolomite interval. The salt deposits could have formed there and been later removed by erosion or dissolution. In fact, the Late Ordovician through Devonian rock record of eastern Utah is relatively thin and ridden with erosional unconformities that signify periods of nondeposition and/or erosion. In

short, in the evaporative reflux model, dolomite is not a primary mineral deposited on the seafloor but instead represents calcium carbonate (limestone) sediments altered after deposition and/or lithification.

There is more. Recent studies of dolomite crusts in modern sabkha environments have suggested that certain microbes can form dolomite as a primary deposit as well. In some of the modern sabkhas of the Persian Gulf area, the slimy substances exuded by sulfate-reducing bacteria can chemically bind calcium and magnesium ions in a manner that results in dolomite crystallization. Indeed, small amounts of dolomite have been identified in the anoxic portions of microbial mats where such bacteria thrive. Some of the Great Basin dolomite may have originated in a similar manner as primary precipitates in the biofilms secreted by microbial organisms. Microbial laminations and stromatolites have been identified in the Laketown and Sevy Dolomites of the Great Basin, so it is at least plausible that microbes may have played a role in their formation. Of course, the concepts of primary dolomite of microbial origin and secondary dolomitization from evaporative reflux are not mutually exclusive: both mechanisms could have been involved during the Great Basin dolomite interval.

Geologists have developed several other explanations for the origin and abundance of dolomite in the prehistoric rock record. Dolomitizing reactions can occur in the mixing zone between subsurface freshwater lenses and saline water migrating landward from the seafloor offshore. Circulation of briny water through carbonate mud driven by geothermal heat could also result in dolomitization, as could the fluids squeezed out of sediments that were compacted by deep burial on the ocean floor. Describing the details of these models would require a book longer than the one you are reading, but all the models suggest that most of the ancient dolomite is produced by secondary conversion from original calcium carbonate minerals. Without dolomitization, these minerals would have been lithified into ordinary limestone.

Whatever mechanisms were involved in the origin of Great Basin dolomite, the overall depositional setting across the miogeocline remained similar from Late Ordovician through Early Devonian time to the shallow marine setting of earlier Paleozoic periods. Dolomitization just happened to be more pervasive during the dolomite interval than at any time before or after due to a unique combination of climatic conditions, paleogeographic location, ocean chemistry, and perhaps microbial activity.

The Great Basin dolomite deposits are most prevalent across the inner miogeocline. In the Silurian and Early Devonian Periods, the miogeocline sloped gently westward, but its extent is unknown because rocks of this age have mostly been removed by subsequent erosion in eastern California and western Nevada. Near the shoreline situated in central Utah, carbonate sediments were deposited in shallow, warm, and perhaps hypersaline water of coastal lagoons and tidal flats. These shallow marine sediments were almost completely dolomitized, probably within a few million years of their deposition. The fine-grained dolomite of formations such as the Sevy and Simonson Dolomites commonly has solution pits, microbial laminations, and stromatolites, all suggesting origin in extremely shallow water. A bit farther offshore, carbonate mud accumulated in a more open marine platform dotted with organic build-ups known as patch reefs. The Laketown and Lone Mountain Dolomites, consisting of nearly pure crystalline dolomite are representative of the outer platform sediments near the western margin of the dolomite tract. Water depth across the inner miogeocline was very shallow, and numerous unconformities in the dolomite sequences indicate periods of exposure likely related to eustatic oscillations.

In central Nevada, west of the near-shore dolomite tract, limestone-dominated sequences such as the Roberts Mountains Formation represent carbonate sediments that escaped the dolomitization process. These deposits accumulated in somewhat deeper water of the carbonate slope offshore of the shallow and gently sloping miogeocline platform. In these murky offshore basins, fine-grained, silty limestone with chert nodules and stringers is the dominant lithology. This pattern of deposition across the Great Basin region suggests that the Great Basin seafloor experienced more subsidence offshore than in the shallow coastal region to the east in Silurian time. This sinking of the outer seafloor tilted the carbonate platform seaward. The tilting facilitated the movement of dolomitizing fluids, caused water depth to increase, and probably created more varied ecological conditions that supported a diverse community of marine organisms.

Silurian and Early Devonian Life on the Great Basin Seafloor

The Silurian and Early Devonian dolomites of the Great Basin are generally less fossiliferous than other marine sediments because the recrystallization of primary calcite during the dolomitization process generally obscures or obliterates organic remains. However, enough fossils have been recovered from the dolomite strata to give us a reasonably good hint of how near-shore marine life evolved after the end-Ordovician extinction. The limestone sequences that accumulated west of the dolomite tract yield abundant and well-preserved fossils that document several evolutionary trends among the invertebrate lineages of the Paleozoic fauna. In particular, the corals and brachiopods are the two groups of marine invertebrates best represented by fossils from the Great Basin dolomite interval. In addition, Late Devonian dolomitic rocks have also produced abundant vertebrate remains documenting a curious horde of early fish that darted through the warm water above the Great Basin seafloor some 400 million years ago.

Corals (phylum Cnidaria, class Anthozoa) suffered a major extinction at the end of the Ordovician Period, but the survivors rebounded in Silurian and Devonian time in the warm shallow seas that extended across the miogeocline. Rapid evolution in several lineages of both solitary and colonial corals resulted in a rich coral fauna that has been documented in the Great Basin (Figures 5.9, 5.10). *Halysites*, commonly known as the "chain coral," is one of the most characteristic fossils of the Laketown Dolomite, Lone Mountain Dolomite, and other Silurian deposits of the great Basin (Figure 5.9). This coral constructed loose colonies consisting of many individual pipe-like coralites with elliptical cross-sections. Each tubular coralite housed a coral animal (polyp), but perforations in the walls allowed the flow of water and nutrients through the entire colony. Viewed from above, the linked elliptical tubes of *Halysites* colonies have the appearance of a chain (Figure 5.9).

Favosites is another genus of colonial corals that is common in Silurian and Devonian strata of the Great Basin. These coral colonies typically consist of many closely packed coralites of more-or-less polygonal shape (Figure 5.10, lower right). Both *Halysites* and *Favosites* are tabulate corals, named for the presence of tabulae (sing. tabula), horizontal plates that are secreted as a floor in the space occupied by the living polyps. As the polyps grew and the colony became large, a new tabula was secreted to separate the coral from its earlier living spaces.

Not all corals of the Silurian and Devonian Periods were colonial. Solitary corals belonging to the Order Rugosa secreted a conical calcite shell, divided internally by numerous longitudinal septa (Figure 5.10, upper left). The curved conical shells often resemble a cow's horn, so the term "horn coral" is commonly applied to these solitary anthozoans. Such coral fossils can be several inches in diameter at the widest end, indicating that the Paleozoic polyps were much larger than their tiny living relatives that build modern coral reefs. The Silurian and Devonian corals very likely possessed stinging cells, as do modern corals, and probably fed on small zooplankton, invertebrate larvae, and possibly swimming organisms. The Silurian and Devonian seas were evidently full of such fodder for corals because the corals grew in great profusion near the edge of the miogeocline, forming reefs and reef-like build-ups that have been identified in the Laketown, Lone Mountain, and Sevy Dolomites. Studies of the oxygen and carbon isotopes in the carbonate rocks that preserve the reefs suggest that they likely formed in water as warm as 30°C (86°F!). Thus, seas that covered the Great Basin during portions of the Silurian Period were warm and tropical, crystal clear offshore, and well agitated due to waves and currents. These conditions appear to have been optimal for the growth of many types of corals. But corals were not the only organisms that contributed to the growth of Silurian and Devonian reefs.

The fossils of a peculiar group of extinct calcareous sponges known as stromatoporoids are sometimes incredibly abundant in the Silurian and Devonian strata of the Great Basin (Figure 5.11). Unlike living sponges, stromatoporoids

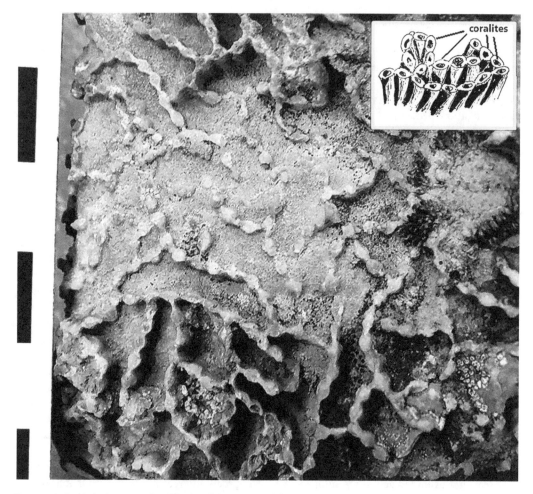

FIGURE 5.9. *Halysites* sp., the Silurian "chain coral," from the Laketown Dolomite of western Utah. Inset sketch illustrates the shape of the loose colony. Bars on the left = 1 cm. Utah Museum of Natural History specimen 4102.

supported their soft tissues with a calcite skeleton that consisted of closely spaced layers connected to each other by tiny pillars (Figure 5.11A). This architecture resulted in a durable boxwork of calcite to resist the seafloor waves and currents. The living sponge cells were probably confined to the outer layers, but as the colony grew the hard structure from earlier growth stages was preserved. The size and shape of stromatoporoid colonies were highly variable, depending on conditions such as water agitation, clarity, temperature, chemistry, and presence of predators. Sometimes, the stromatoporoids would form a wavy layered mass similar in appearance to calcified microbial mats. In calm and shallow water near the coast, a twig-like branching form known as *Amphipora* was so abundant that stromatoporoid thickets carpeted the floor of lagoons and coastal embayments.

Eventually, the branching sponge colonies broke apart, leaving tangled heaps of tubular fragments on the seafloor that were later buried in carbonate mud. For nearly a century, geologists have used the informal term "spaghetti rock" for Silurian and Devonian deposits with such concentrations of dismembered branching stromatoporoids (Figure 5.11B). Several

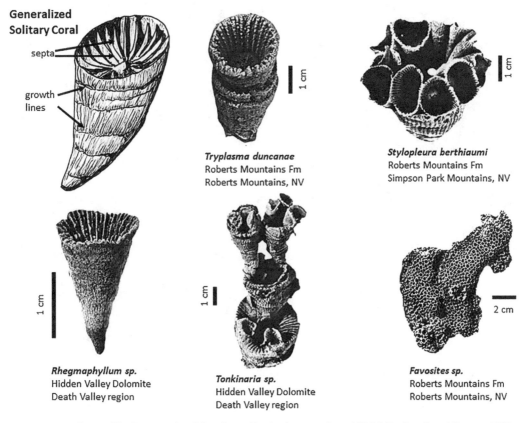

Generalized Solitary Coral

septa

growth lines

Tryplasma duncanae
Roberts Mountains Fm
Roberts Mountains, NV

1 cm

Stylopleura berthiaumi
Roberts Mountains Fm
Simpson Park Mountains, NV

1 cm

Rhegmaphyllum sp.
Hidden Valley Dolomite
Death Valley region

1 cm

Tonkinaria sp.
Hidden Valley Dolomite
Death Valley region

1 cm

Favosites sp.
Roberts Mountains Fm
Roberts Mountains, NV

2 cm

FIGURE 5.10. Some Silurian corals of the Great Basin. Images from USGS Professional Papers 777 and 973, listed in Chapter 5 references.

"species" of *Amphipora* have been identified by paleontologists, each distinguished by size and pattern and branching. Some stromatoporoid colonies were irregular encrusting forms that grew on, and sometimes covered, objects on the seafloor including corals and shells of other invertebrates such as brachiopods or bryozoans. Regardless of the shape of the colonies, stromatoporoids played a major role in constructing the reefs on the Great Basin seafloor during the Silurian and Devonian Periods (Figure 5.11C). In addition to corals and stromatoporoids, Great Basin dolomite strata sometimes preserve hemispherical or bulbous stromatolites, indicating that microbial colonies also contributed to reef construction. The coral-stromatoporoid reef communities are unique to the Silurian-Devonian slice of geologic time. What is most notable is that at no other time during the in-

credibly long history of the Great Basin seafloor were reefs constructed by this same biotic association.

Brachiopods are also common as fossils in the Silurian and Devonian rocks of the Great Basin (Figure 5.12), but they too tend to be most common and best preserved in limestone deposits that accumulated offshore from the dolomite tract. In central Nevada, for example, brachiopods are among the most abundant fossils in such formations as the Rabbit Hill Limestone, McColley Canyon Formation, and Denay Limestone. The Lone Mountain and Laketown Dolomite formations also yield sufficiently abundant and well-preserved brachiopod fossils to be useful for correlation and faunal studies. Overall, the Silurian and Devonian brachiopod faunas of the Great Basin are comprised of many lineages that survived the end-Ordovician

FIGURE 5.11. Devonian stromatoporoid fossils from the Great Basin: (A) hemispherical stromatoporoid from the Guilmette Limestone, West Pahranagat Range, Nevada; (B) *Amphipora*, a twig-like stromatoporoid from the Simonson Dolomite, Hiko Range, Nevada; (C) a stromatoporoid reef in the Guilmette Limestone, Hiko Range, Nevada.

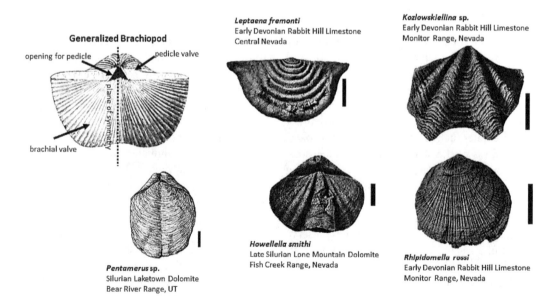

FIGURE 5.12. Generalized brachiopod shell anatomy (*upper left*) and some selected brachiopods from Silurian and Devonian strata of the Great Basin. Images from USGS Professional Papers 777 and 973, listed in Chapter 5 references.

extinction, spread across the Great Basin sea-floor as conditions stabilized, and diversified into many new types adapted to a wide variety of marine habitats. On a global scale, the brachio-pods became one of the most prominent com-ponents of the Paleozoic fauna by the middle part of the Devonian Period. Brachiopods seem to have been particularly well adapted to the coral-stromatoporoid reef habitat. Paleontolo-gists have identified 46 different associations of brachiopods in the Early Devonian alone, each representing a community adapted to specific microenvironments on the miogeocline.

Several other groups of organisms are also prominent, if not dominant, in the coral-stromatoporoid-brachiopod fossil assemblages from Silurian and Devonian strata. Trilobites, cephalopods, gastropods, crinoids, bryozoans, echinoids, and conodonts all rebounded from the end-Ordovician crisis and are important elements of the general fauna known from Silu-rian and Devonian strata. Among the cephalo-pods, a new group of swimming predators, the ammonoids (class Cephalopods, subclass Ammonoidea), evolved a coiled shell near the end of Silurian time similar in appearance to the modern nautilus. By the beginning of the Devonian Period, about 420 million years ago, there were more than 300 families of marine invertebrates in the Paleozoic fauna, marking a full recovery of the biodiversity lost during the end-Ordovician extinction.

The Devonian also heralded an age of re-markable evolution among a group of sea crea-tures that had previously occupied only minor positions in the ecological structure of marine life on the Great Basin seafloor. Soon after Brit-ish geologists established the Devonian Period in 1839, study of fossils from rocks of this age revealed a burst of evolution among marine vertebrates marking the origin and rapid pro-liferation of many different lineages of fish and fish-like creatures. From the late 1800s on, the Devonian was referred to as the Age of Fishes, an informal epithet that conveys this global surge. While the earliest fish-like vertebrates likely evolved in Cambrian time, it was in the Devonian Period that the seas of the Great Basin seafloor filled with a remarkably diverse fauna.

Many types of fish—including some of the most bizarre species to ever swim the seas—cruised the Silurian and Devonian reefs.

Silurian and Devonian Fish of the Great Basin

In modern seas, the vertebrates we call fish can be easily assigned to one of the three categories of the chordate subphylum Vertebrata recog-nized by scientists:

1. jawless fish of the superclass *Agnatha*: this group includes the modern lampreys and hagfish;
2. jawed fish with cartilaginous skeletons of the superclass Gnathostomata, class Chon-drichthyes: the sharks, rays, and skates;
3. jawed fish with bone skeletons of the super-class *Gnathostomata*, class *Osteichthyes*: all familiar fish such as salmon, cod, bass, etc.

Each of these groups is, of course, subdivided into numerous subordinate orders and families. While we will avoid most of the complexities of fish classification, to understand the variety of Paleozoic fish in the Great Basin, we need to introduce a few of the subcategories and iden-tify some of their extinct relatives. These groups appear in the table summarizing the occurrence of Devonian fish in the Great Basin (Figure 5.13).

The jawless agnathan fish include two extinct groups of armored fish that flourished in the Early and Middle Paleozoic seas: the osteo-stracans (order *Osteostraci*) and the hetero-stracans (order *Heterostraci*). The heterostracans were extremely primitive, lacking fins other than the tail. The head region was enclosed in a bony carapace, and the body had thick bone scales of distinctive form and structure. The heterostracans first appeared in the Silurian Period and were probably slow-moving bot-tom feeders consuming algae, dead organisms, and other organic matter. Osteostracan fish were relatively advanced agnathans, charac-terized by bony shields and plates around the head, flexible scales covering the body, and powerful tails. With paired, spiky fins and a torpedo-shaped body, the osteostracans were likely active swimmers that cruised the shal-low seafloor consuming soft food. At least some

CHAPTER 5

Taxonomic Group		Lost Burro Formation Death Valley region	Red Hill Beds Central Nevada	Denay Limestone Roberts Mts., Nevada	Sevy Dolomite Western Utah & Nevada	Water Canyon Formation Northern Utah
AGNATHA	Osteostraci	Cephalaspis			Cephalaspis	Cephalaspis, at least 3 sp.
AGNATHA	Heterostraci	Blieckaspis Panamintaspis several other undescribed taxa unnamed actinolepid			Tuberculaspis Lamiaspis Pirumaspis Ctenaspis	Oreaspis, at least 3 sp. Blieckaspis Clydonaspis Pirumaspis Cosmaspis Cordipeltis Allocryptaspis, 2 sp.
GNATHOSTOMATA	Placodermi	Onchus Dunkleosteus	Asterolepis	Asterolepis	Aleosteus	Simblaspis Asterolepis Aethaspis, 2 sp. Bryantolepis
GNATHOSTOMATA	Acanthodii		Machaeracanthus Persacanthus	Persacanthus Onchus Nostolepis (McColley Cyn Fm)	Onchus Sevyacanthus Nodocosta	Onchus sp. Cacheoacanthus Ischnacanthus Ptychodictyon
GNATHOSTOMATA	Chondrichthyes	unnamed cladodont shark cochliodont teeth	Polymerolepis	Petalodontida		
GNATHOSTOMATA	Osteichthyes		Cheirolepis Eusthenopteron, a sarcopterygian Tinirau, a sarcopertygian 2 dipnoans	Cheirolepis		Dipterus sp., a dipnoan

FIGURE 5.13. A partial list of fossil fish genera described from Silurian and Devonian rocks of the Great Basin. This table is incomplete and only intended to illustrate the diversity and abundance of fish fossils in Great Basin strata.

osteostracan fish appear to have had advanced sensory organs used to locate prey or sense the presence of predators. Both heterostracans and osteostracans became extinct at the end of the Devonian Period.

The Gnathostomata, or jawed fishes, are much more varied and abundant in the modern world than the agnathans. As we have already seen, the two primary groups of modern gnathostomes are the cartilaginous fish (sharks, etc.) and the bony fish. Both groups are represented in the Silurian and Devonian fossils from the Great Basin. However, fossils also document the presence of several other extinct groups of jawed fish in the Great Basin seas. The placoderms (class Placodermi) were a large and varied group of armored fish that included predators and scavenging bottom feeders. In general, the placoderms possessed a mosaic of articulating body plates armoring the head region, hinged jaws with teeth or sharp bony blades, and multiple sets of paired fins supported by spikes of bone. One group of predatory placoderms, the arthrodires (order Arthodira), had a joint con-

necting the top of the armored head to the back of the skull and the body. This unique articulation allowed the mouth to open into a huge cavity capable of engulfing large prey. Lined with sharp and serrated bone ridges, the powerful jaws could crush, slice, and dice even well-armored prey. Some of the arthrodires grew to a length of 20 feet and must have been fearsome superpredators in the Paleozoic seas, similar in ecological niche to the modern Great White shark. Many paleontologists consider the placoderms to be a "polyphyletic" group. That is, this category of fish includes forms that may have descended from different ancestors and, though similar in character, are not all closely related.

Another probable polyphyletic group of prehistoric fish prevalent among the Great Basin fossils is the acanthodians (order Acanthodii). The acanthodians were relatively small, usually only a few inches long, with characteristics of both bony and cartilaginous fish. These fish were shark-like, with a slender streamlined body, paired pectoral and pelvic fins, dorsal fins supported by robust spikes of bone, and an upturned tail. Acanthodian skeletons were mostly cartilage, but the other characteristics, such as their V-shaped gill supports, are more like the osteichthyan fish. The body was covered by pavement of close-fitting, thick rhombohedral, polygonal, or blocky scales composed of durable enamel and bone-like layers. These robust scales and the stout bony spines that supported the fins are commonly found as isolated fossils in Great Basin Devonian strata.

Several different types of shark-like cartilaginous fish have also been identified. Though these fossils are clear evidence of a diverse shark fauna, nearly all consist of tiny scales, teeth, tooth plates, and scraps of ossified cartilage. Such fragmentary remains make it difficult to know much about the morphology, ecology, and precise relationships among the various sharks that cruised the shallows of the reefs and lagoons across the miogeocline. However, the teeth do indicate that some were active predators and likely agile swimmers while others probably swam slowly over the bottom, crushing shelled organisms with broad, plate-like teeth. None of the sharks were large enough to rival the arthrodire placoderms as the top predator in the Devonian seas of the Great Basin region.

Though the bony fish (class Osteichthyes) dominate modern marine fish communities, they are relatively uncommon components of the Great Basin Devonian assemblages. But two groups are of particular interest because they signify early adaptations of some fish lineages to terrestrial life. These adaptations ultimately led to the emergence of tetrapods capable of colonizing land habitats. Because complex vascular plants first invaded dry land in the Devonian, it is not surprising that the riot of fish evolution in this same period includes a path leading to tetrapods. The greening of terrestrial environments by plants in the Devonian signifies that land habitats were no longer as hostile to life as they had been through most of the earlier Paleozoic era. The movement of plant life from the oceans to land created an ecological opportunity never before available to vertebrates. Then, as now, such ecological voids do not remain vacant for

long. By the end of the Devonian Period, the first terrestrial vertebrates were beginning to crawl ashore.

The two osteichthyan groups that led the way to land both belong to a subcategory known as the sarcopterygians (technically the Sarcopterygii, variously considered a class, subclass, or clade by biologists). All sarcopterygians are commonly described as lobe-finned fish because their paired fins are connected to the body via a fleshy, muscular lobe or stalk. The lobes were anchored to the body by bones that allowed propulsion by the limbs and fins rather than the tail. This unique structure allowed the sarcopterygians to lift and push themselves over land at least occasionally. Some sarcopterygians also possessed rudimentary nostrils and lungs to supplement their gills, a further indication of their adaptation to land environments. Lungfish (subclass Dipnoi) are surviving sarcopterygians descended from ancestors that first appeared in the Devonian Period. The other Devonian fish that crossed the threshold to land are classified within the Tetrapodomorpha, a group that includes the ancestors of the modern amphibians and reptiles. Both types of sarcopterygians are represented in the Devonian fish faunas of the Great Basin.

Isolated fragments of bony material have been discovered in Ordovician strata of the Great Basin, but the fossils are too fragmentary to fully determine what types of fish evolved from their Cambrian chordate ancestors. In particular, the Ordovician Swan Peak Formation of north-central Utah has produced bony plates and fragments that likely represent the armor of early osteostracans. By Silurian time, several different lineages became established in the Great Basin seas, including the more advanced jawed acanthodians and placoderms. Still, the fossils of these fish are mostly small and dissociated elements such as scales, fin spines, and jaw fragments. These have been reported from the Roberts Mountains Formation of central Nevada.

Not surprisingly, the Great Basin Devonian fossil record also documents an extraordinary explosion in the diversity and abundance of

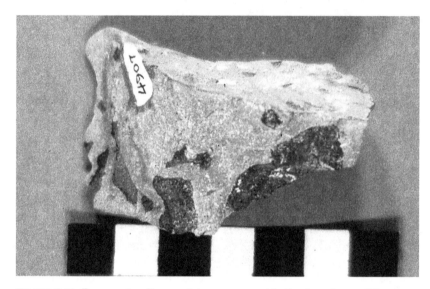

FIGURE 5.14. Fragments of bony plates preserved in the limestone of the Devonian Water Canyon Formation of the Lakeside Mountains, Utah. Scale bar divisions = 1 cm.

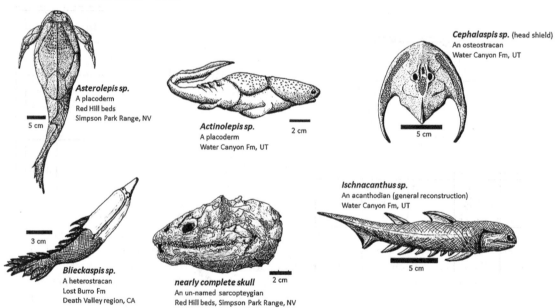

Cephalaspis sp. (head shield)
An osteostracan
Water Canyon Fm, UT

Asterolepis sp.
A placoderm
Red Hill beds
Simpson Park Range, NV

5 cm

Actinolepis sp.
A placoderm
Water Canyon Fm, UT

2 cm

5 cm

Ischnacanthus sp.
An acanthodian (general reconstruction)
Water Canyon Fm, UT

3 cm

5 cm

Blieckaspis sp.
A heterostracan
Lost Burro Fm
Death Valley region, CA

nearly complete skull
An un-named sarcopteygian
Red Hill beds, Simpson Park Range, NV

2 cm

FIGURE 5.15. Examples of Devonian fish of the Great Basin. These taxa are only a few of the many fish identified in the Devonian strata.

Paleozoic fish. Fossils of several different fish lineages have been recovered from Devonian formations across the region, from the Death Valley to north-central Utah (Figure 5.13). Fragments of skeletal material and bony armor plates are sometimes as common as brachiopods, corals, and sponges in rocks of Middle to Late Devonian age (Figure 5.14). Often, these bits of bony material are too fragmentary to identify to the genus and species level, but their abundance clearly indicates the presence of a diverse and extensive fish fauna. Some Devonian fossil localities in the Great Basin have produced spectacularly abundant and uncommonly well-preserved remains of ancient fish. This includes Red Hill in Nevada's northern Simpson Park Range where hundreds of well-preserved fish fossils have been found, along with remains of sponges, brachiopods, gastropods, conodonts, and corals. The fossils at Red Hill occur in platy slabs of thin-bedded, silty limestone deposited in a complex system of reefs, lagoons, and bays 50–100 miles offshore. The Water Canyon Formation of north-central Utah is equally famous for Devonian fish fossils. It is dominated by sandy and silty limestone deposited in near-land coastal settings. Together, the Red Hill beds and the Water Canyon strata provide an excellent view of the many types of fish that swam the Great Basin reefs, coastal bays, and estuaries during the Devonian Period.

Placoderms, acanthodians, cartilaginous fish, heterostracans, and osteostracans have all been identified from the fish fossils collected from Devonian rocks of the Great Basin (Figure 5.15). Most of these fish were relatively small, rarely larger than a few inches long, comparatively poor swimmers, and filled a variety of ecologic niches including bottom-feeding scavengers, microbial grazers, and suspension-feeders. Among the Devonian fish are several types of placoderms, such as *Actinolepis* and *Dunkeleosteus*, that evince the presence of larger, more heavily armored predators in the Devonian seas. Several types of acanthodians, with their torpedo-shaped bodies and large spiky fins, were small but active swimmers among the reefs, lagoons, and inlets of the eastern Great Basin. Ray-finned bony fish (class Osteichthyes, subclass Actinopterygii) were also present. One of the best known of these is *Cheirolepis*, an active predator with triangular scales, large eyes, and tooth-lined jaws that swept the near-shore shallow waters. *Cheirolepis* was large—nearly two feet long–and evidently could tolerate the brackish and fresh water near land.

Remains of several different lobe-finned sarcopterygian fish have also been discovered at Red Hill and in the Water Canyon Formation. These fish were the first vertebrates in the Great Basin to wiggle ashore—at least tempo-rarily—from the seas covering the Great Basin seafloor. Devonian sarcopterygians include both lungfish (dipnoans) and stem tetrapods such as *Tinirau clackae* (Figure 5.16). Adaptations for terrestrial life in these groups, such as fleshy fin supports, robust shoulder and pelvic bones, and paired secondary nostrils leading to rudimentary lungs, all suggest these fish were ancestors of the earliest terrestrial vertebrates. The transition of vertebrates from marine and aquatic habitats to dry land is one of the great evolutionary milestones in the history of life on our planet. No one can say when and where the first land-living vertebrate came ashore. Most likely, it took place in many locations at various times. But fossils from the Great Basin clearly demonstrate that "fish" were about to cross that ecological threshold in the shallows of the Great Basin seafloor by Late Devonian time.

The Late Devonian Extinctions

Over the final 15 million years of the Devonian Period, from about 375 to 360 million years ago, at least three waves of extinction affected marine life on a global scale. The most significant of these extinction events occurred near the boundary between the Frasnian and Famennian ages, the final two increments of Devonian time. For this reason, the end-Devonian extinction is sometimes referred to as the Frasnian/Famennian extinction. This biotic crisis was originally known as the Kellwasser event, named for the locality in Germany where geologists discovered initial evidence in the form of black shales deposited in anoxic marine basins. Some paleontologists group the multiple extinction events together as the singular "late Devonian extinction." Whatever term is used, the great demise of marine biodiversity at the close of Devonian time ranks as one of the five greatest mass extinctions of the Phanerozoic Eon (Figure 4.11). Overall, scientists have estimated that 70–80 percent of all animal species, and some 20 percent of all animal families, disappeared by the end of Devonian time.

The Late Devonian extinctions strongly affected marine life on a global scale, but the effects were more severe on some groups of organisms than others. The reef-building organisms, such

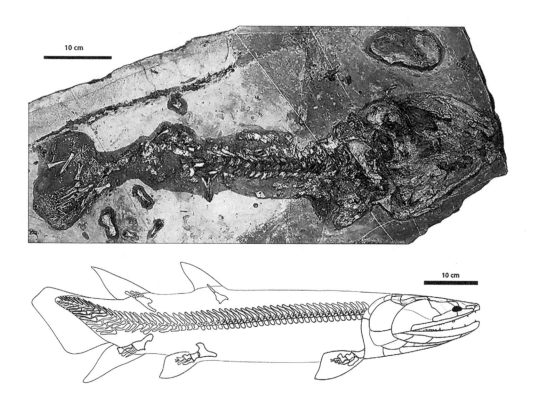

FIGURE 5.16. Nearly complete skeleton (*top*) and reconstruction of the stem tetrapod *Tinirau clackae* from the Devonian Red Hill beds, central Nevada. Figures from Swartz (2012).

as the stromatoporoids, corals, and sponges, suffered the greatest effects. The collapse of reef communities also strongly affected the several lineages of brachiopods, trilobites, ammonoid cephalopods, and conodonts that were well adapted to the warm tropical seas. Among the fish, the primary victims of extinction were the agnathans, placoderms, acanthodians, and some sarcopterygians. Oddly, the cartilaginous fish and the ray-finned actinopterygians continued to flourish. Generally, it was the organisms best adapted to warm, shallow, tropical seas that were the most vulnerable.

As is almost always the case in mass extinction events, no single source of biotic stress seems responsible here. Instead, there appear to have been multiple simultaneous disruptions to the marine ecosystem, resulting in a prolonged ecological crisis. The Late Devonian rock record suggests a highly unstable sea level during that time. In the western United States,

no fewer than 18 significant fluctuations in sea level have been interpreted from the sedimentation patterns. Abrupt sea-level drops leading to multiple severe regressions, seem to alternate with transgressions resulting from periods of eustatic rise. The roller coaster of Late Devonian eustatic changes may be linked to climatic deterioration and southern hemisphere glaciation, to tectonic changes in ocean basins, or to landmass reconfiguration. Moreover, the eustatic oscillations were superimposed on a general decline of sea level that signified the end of the Tippecanoe sequence (Figure 4.2). Regardless of their origin, these vacillations caused repetitive changes in water depth, clarity, temperature, and circulation that undoubtedly resulted in biotic stresses on shallow marine organisms across the Great Basin seafloor.

Along with the eustatic instability, there were several other geological, oceanographic, and biological events that could have played a role in the Late Devonian extinctions. Extensive black shale deposits formed in Late Devonian basins worldwide. Rich in residual organic matter, they indicate poorly oxygenated bottom waters and a stagnation of oceanic circulation. This stagnation is further supported by numerous positive $\delta^{13}C$ excursions detected in the latest Devonian strata of the Great Basin. This enrichment of carbon-13 is generally thought to result from burial of large amounts of undecomposed organic matter on the seafloor. Also, continental greening from the invasion of land by large vascular plants in the Late Devonian may have drawn down atmospheric carbon dioxide enough to initiate a general cooling trend. Land plants also accelerated the weathering of rock and formation of soils, processes that further withdraw carbon dioxide from the atmosphere. Additionally, there was also a burst in igneous activity, particularly in modern Russia, Siberia, and Australia about the same time as the Late Devonian extinctions were in progress. The eruption of so much lava may have released enough ash and gases into the atmosphere to have had global effects. Finally, there is widespread geological evidence (including some from the Great Basin, as we will see in the next chapter) that a storm of comets or asteroids lasting several million years may have influenced global climate, sea level, and ocean chemistry.

Regardless of these possible factors, once the extinctions finally played out, the survivors would repopulate the Great Basin seafloor but with a different configuration of species and a distinctive ecological structure. But even as the Devonian extinctions were reshaping marine life in the ancient seas, the deep foundation of the Great Basin seafloor was itself rumbling to life. After some 340 million years of passive margin serenity, tectonic turmoil in the Late Devonian Period began to restructure the western edge of Laurentia. New plate tectonic interactions generated powerful earthquakes that shook the seafloor and crushing geological forces tore across the region. Sediments that had accumulated in murky basins offshore of the shallow carbonate platform were twisted, torn, and tortured into mangled heaps of rock driven from the depths to emerge as land near the edge of the formerly submerged miogeocline. This major mountain-building event, the Antler orogeny, began in Late Devonian time and continued for millions of years well into the succeeding Mississippian subperiod and beyond. While the Antler orogeny was in progress, the comet storms continued, including a very significant event, the Alamo impact, in southern Nevada.

Shaken from below and bombarded from above, the Great Basin seafloor would never be the same after the turmoil that began in Late Devonian time. In fact, the transition from geological serenity to tectonic tumult ultimately led to the demise of the Great Basin seafloor itself.

6

The Devonian-Mississippian Interval

Tectonic Tumult and Cosmic Calamities

In the 1950s, before the development of modern analytical tools and the plate tectonic theoretical framework, geologists working in the Great Basin were occupied primarily with mapping and correlating rock formations across the relatively unexplored region. In the same era, the mining of rich ore deposits in many Basin locations stimulated great interest in unraveling the region's geological structure. Legions of geologists swarmed the region, and their work established many of the fundamental precepts and geological patterns we have explored in previous chapters. During the 1950s and 1960s, geologists first noticed evidence in the rock record that something very interesting happened across the Great Basin seafloor beginning in the Late Devonian. The decades of research that followed have confirmed that the Late Devonian–Mississippian interval was a time of fascinating, though still not completely understood, geological upheavals along the western margin of Laurentia.

Geologists working in the highly mineralized Battle Mountain region of central Nevada, first noticed evidence for some sort of geological Late Devonian ruckus. Mangled masses of Early Paleozoic deep-water marine rocks seemed to be stacked on top of each other in great sheets resembling shingles on a roof, with several sheets of Cambrian through Devonian deep-water sediments emplaced above shallow-water deposits of similar age. The deep-water succession was dominated by organic-rich shale, chert, and bedded barite (a barium sulfate mineral) deposits that had accumulated in the eugeocline of western Nevada. Such rocks comprise the mostly Ordovician Vinini and Valmy Formations, as well as the younger Elder Sandstone and the Slavern Chert. They were perched above a limestone and dolomite succession of the Early-to-Middle Paleozoic Age that accumulated in the shallow water of the miogeocline in central Nevada. The surfaces separating the two assemblages were mapped as low-angle thrust faults, though they were generally not well exposed. Above the inferred thrust faults, the transported rocks were folded, twisted, and shattered.

This exceedingly complex geological structure could be interpreted in several ways. The original interpretation, still considered valid by most geologists, was that each sheet of deep-water marine strata represented a slab of rock forced eastward along thrust faults from the deep eugeocline, up the continental slope, and over the outer edge of the shallow miogeocline platform. The principal thrust faults (or simply "thrusts," an abbreviation commonly used by geologists) separating each slab were named, mapped, and traced to other areas of northern Nevada. Because the thrust sheets carried strata of Cambrian through Middle Devonian age, and the youngest rocks overridden by the plates were also Middle Devonian, it was surmised that the faulting began sometime in the Late Devonian Period and likely continued

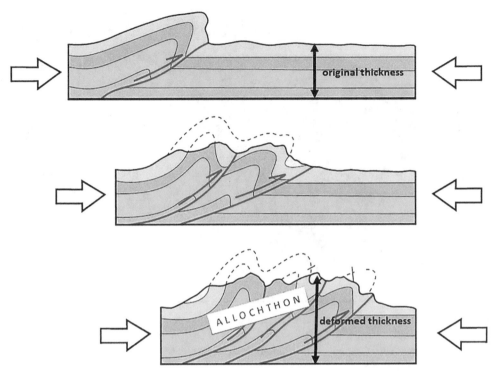

FIGURE 6.1. The geometry of thrust faults, or simply "thrusts," requires compression and results in the overall shortening and thickening of the rock succession. The entire mass of mangled rock transported along a system of imbricate thrusts is known as an allochthon. Background graphic from DeCourten and Biggar (2017), listed in Chapter 1 references.

into succeeding periods. Several different thrust faults were named from localities in northern Nevada where they were originally mapped. The most extensive and best studied of these was the Roberts Mountains thrust, named for the prominent mountain range in Eureka County, north-central Nevada.

The geometry of thrust faults results from a mass of rock situated above the fault plane moving up and over the mass of rock beneath it. This requires the application of compression, or squeezing forces, as opposed to other types of faults that form from tension or shear. Not surprisingly, a thrust fault tends to shorten the earth's crust by displacing one slab of rock over another. Typically, the mass of rock above the fault plane is folded and twisted during transport, and this folding shortens the crust even more. When several thrust faults develop within a stack of rock strata, the shingle-like structure

is described as imbricate thrust sheets, and the overall shortening and thickening of the crust can be very significant. The total mass of rock transported, deformed, and piled above the lowest thrust is called an allochthon: a mass of deformed rock that has been tectonically moved from its place of origin (Figure 6.1).

The identification of Middle Paleozoic thrust faults in the Great Basin signifies the transformation of the passive margin of western Laurentia into a more geologically active plate tectonic setting. The compression necessary to activate the Middle Paleozoic thrusts, deform the strata in complex folds, and transport the allochthons eastward, would have been impossible without the convergence of lithospheric plates somewhere nearby. For nearly 400 million years, such interactions had not disturbed the geological tranquility of the Great Basin seafloor. In Late Devonian time, and continuing

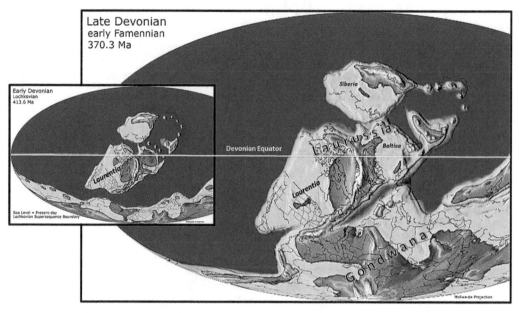

FIGURE 6.2. Paleogeographic reconstructions of Early Devonian (*inset*) and Late Devonian time (*background*). Images from C.R. Scotese (2014). Atlas of Devonian Paleogeographic Maps, PALEOMAP Atlas for ArcGIS, vol. 4, The Late Paleozoic, Maps 65–72, Mollweide Projection, PALEOMAP Project, Evanston, IL.

well into the following Mississippian subperiod, all that changed. Convergent plate interactions seem to have developed in western Laurentia, and the compressive forces they generated raged across the Great Basin region to activate numerous thrusts and transport allochthons eastward from Idaho to southeast California. We now know that this deformation was just the beginning of a geological melee that would last for more than 100 million years.

What could have caused the transformation of the passive margin of Laurentia into a tectonically active, presumably convergent, plate boundary? This transition is likely linked to global changes in plate configurations that took place about the same time as the Middle Paleozoic deformation in the Great Basin began. Prior to Late Devonian time, Laurentia was a mostly isolated continent straddling the equator (Figure 6.2). Earlier terrane collisions on the eastern margin of the continent had raised the Taconic Mountains, but the western edge of Laurentia remained a passive margin. In Late Devonian time, major orogenic events resulting from col-

liding landmasses on the eastern and northern sides of Laurentia accompanied the piecemeal construction of immense supercontinents. The Calendonian orogeny of the British Isles, Scandinavia, and Siberia culminated when large masses of crust known as the Baltica and Siberia terranes crashed into Laurentia (Figure 6.2). From Middle Devonian to Early Mississippian time, a smaller mass known as Avalonia collided with the northeast edge of Laurentia, resulting in the Acadian orogeny of the northern Appalachian Mountains. Through these tectonic events, Laurentia became so enlarged that—after Late Devonian time—some geologists refer to it as Laurussia, a much larger continent than was Laurentia. Laurussia was one of the two great global supercontinents to exist in later periods of the Paleozoic Era. The trend of continental amalgamation continued until the dawn of Mesozoic time, 250 million years ago. At that time, the suturing of Laurussia to Gondwana, the other Late Paleozoic supercontinent, would form Pangaea, the most recent supercontinent in earth history. Surrounded by a single, immense global ocean, Pangaea contained virtually all the planet's land areas.

Compared to all the major mountain-building events taking place around the world in Late Devonian-Mississippian time, the geological fracas in the Great Basin was a relatively small disturbance. Nonetheless, it may represent an expression of the buttressing effect from so much land being added to the northern and eastern sides of Laurentia. Presumably, seafloor spreading was still active far offshore of Laurentia's western edge. Due to collisions on its eastern flank, the continent was no longer able to move freely; the oceanic crust along the passive western margin may have buckled, with a portion forced downward in a subduction zone somewhere near the Great Basin region. This event initiated a convergent plate boundary an unknown distance offshore of western Laurentia (or Laurussia). Driven by the compression of converging plates, the Late Devonian-Mississippian thrust faults signify the transformation of the passive margin to an active plate boundary. Rocks were crushed and shoved against the edge of the continent, earthquakes shook the ocean floor, and mountainous highlands rose from the sea. The Great Basin seafloor's long period of tectonic tranquility had ended.

According to the classical view of Middle Paleozoic mountain-building in the Great Basin, numerous allochthons underlain by the Roberts Mountains thrust were stacked together along the outer edge of the shallow miogeocline. Prior to later erosion, the total thickness of the imbricate thrust sheets may have been as much as 16,000 feet (5 km). This huge mass of deformed sediments was tectonically transported eastward some 50–75 miles (60–120 km) up the continental rise and slope, ultimately overlapping the western margin of the miogeocline. There, the stack of allochthonous strata significantly thickened the crust by Early Mississippian time (Figure 6.4). When the Earth's crust becomes thickened in this manner, it tends to rise higher above the dense rocks of the lower lithosphere in the same manner that a thick block of wood floats higher in water than a thin sheet of plywood. The uplift resulting from crustal thick-

ening in the Late Devonian created a landmass comprised of deformed and faulted eugeoclinal rocks in the middle of the Great Basin seafloor (Figures 6.3, 6.7). Geologists named this mountain-building event the Antler Orogeny for Antler Peak near Battle Mountain, where the deformation was first identified. The linear mountain system that was lifted, the Antler orogenic belt, was aligned roughly along the outer margin of the miogeocline and stretched southward from at least as far north as Idaho to southeastern California. This major mountain system, the first to rise from the Great Basin seafloor since its inception, would persist well into the Pennsylvanian subperiod.

The basic concept of Middle Paleozoic mountain-building, and the Antler orogeny itself, grew from the interpretation of the contact between the shelf rocks in central Nevada and the overlying deep basin strata as thrust faults. In fact, the thrusts themselves are almost always very poorly exposed, and the fault zones are almost impossible to examine closely in outcrops. This difficulty in affirming the thrust fault interpretation of the contact left open the possibility of other explanations. For example, the abrupt placement of deep basin sediments over shallow marine strata may have been caused by rapid subsidence of the seafloor along high-angle faults that do not require compression. Also, as more areas of the Great Basin were explored, some geological evidence suggested that several thrust faults thought to have formed during the Antler orogeny were much younger, exhibiting movement as late as the Mesozoic Era. Exposures of shallow-water limestone within the tract of deep-water sediments were originally interpreted as "windows" in the thrust sheets, where erosion had removed the cover of eugeoclinal strata to reveal the underlying shallow-water carbonate strata. New evidence (such as that described later in this chapter) suggested that some of these exposures were, in fact, giant blocks of limestone that had been dislodged from the edge of the shelf and slid down the continental slope into the ooze of the deeper ocean. Moreover, the most popular model for the Antler orogeny involved the collision of an

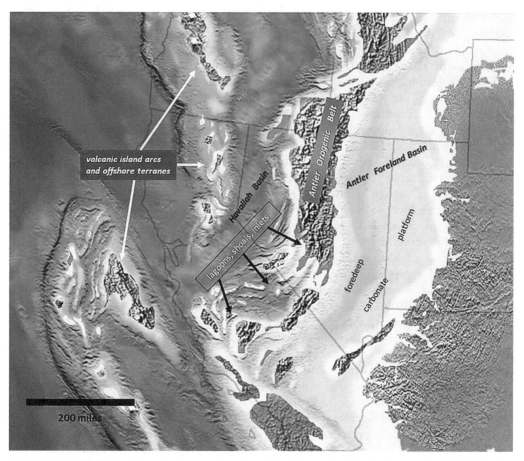

FIGURE 6.3. Late Devonian paleogeography of western Laurentia. The Antler orogenic highland
was raised during the Antler orogeny along the former western margin of the miogeocline. Two
submerged basins flanked the Antler belt: the Antler foreland basin to the east and the Havallah
Basin to the west. Paleogeographic base map prepared by R. Blakey, Colorado Plateau Geosystems Inc.

offshore volcanic island arc with the continental
shelf to initiate the deformation. However, the
volcanic rocks of this island arc, or even remnants of them, could not be located anywhere
in the Great Basin.

These aspects of the Antler orogeny remain
a mystery to the present time. Geologists at-
tempting to understand the plate tectonics of
the Antler orogeny over the past several decades have tried to link the event to a specific
interaction of plates along western Laurentia.
It is a bit beyond our scope to review all the
details of the plate tectonic models that have
been proposed for the Antler orogeny, but the

most prominent ones have involved plate convergence with subduction of an oceanic plate to
the east, to the west, with a volcanic arc, and
without a volcanic arc. Each of these models has
at least some supporting evidence, but all have
some difficulties in explaining the timing, geographic extent, and structural details. Still, there
is general agreement on one simple conclusion:
something happened in the Great Basin near the
end of Devonian time to create land along the
outer edge of the miogeocline. The undeniable
evidence for the uplift is found in the rocks that
formed during and after this disturbance. Whatever the plate tectonic mechanism for the Antler

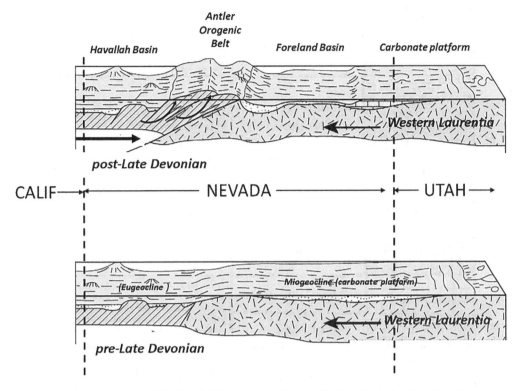

FIGURE 6.4. A general model for the Antler orogeny. Eugeoclinal sediments were thrust eastward toward the edge of the miogeocline. There, the allochthons rose as an exposed chain of deformed sediments, the Antler orogenic belt. This model is one of several possible interpretations of Devonian–Mississippian mountain-building in the Great Basin. Modified from a USGS graphic.

orogeny may have been, plate interactions of some sort clearly began to lift a portion of the Great Basin seafloor above sea level some 360 million years ago.

Sedimentary Record of the Antler Orogeny: The Antler Foreland Basin

Soon after the thrust faults and deformed allochthons characterizing the Antler orogenic belt were first mapped and described, geologists also developed a framework for interpreting the sedimentary rocks that accumulated near the rising mountain chain. In several different Late Devonian through Pennsylvanian rock sequences of the eastern Great Basin, the normal Paleozoic pattern of carbonate deposition was abruptly interrupted by a flood of coarse siliciclastic sediment such as conglomerate and

sandstone (Figure 6.5). These sediments, such as those in the Diamond Peak Formation of east-central Nevada, were primarily comprised of rock particles derived from the bedrock of the Antler orogenic belt (Figure 6.6). The coarse-grained conglomerate and pebbly sandstones were interpreted as debris created by the vigorous erosion of the rising Antler orogenic belt. This erosion left a prominent unconformity within the Antler belt, separating the rocks of lifted allochthon and much younger overlying strata. Adjacent to the Antler orogenic belt, flanking basins subsided rapidly during and after the Antler orogeny, allowing thick sequences of erosional refuse to accumulate on the sinking seafloor. The main area of sediment accumulation was an elongate trough located immediately east of the Antler orogenic belt that

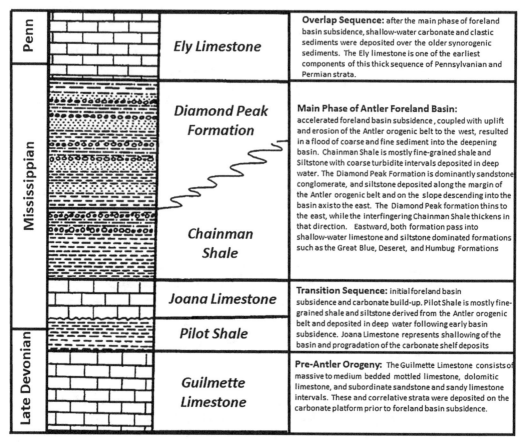

			Overlap Sequence: after the main phase of foreland basin subsidence, shallow-water carbonate and clastic sediments were deposited over the older synorogenic sediments. The Ely limestone is one of the earliest components of this thick sequence of Pennsylvanian and Permian strata.
Penn		Ely Limestone	
Mississippian		Diamond Peak Formation Chainman Shale	Main Phase of Antler Foreland Basin: accelerated foreland basin subsidence, coupled with uplift and erosion of the Antler orogenic belt to the west, resulted in a flood of coarse and fine sediment into the deepening basin. Chainman Shale is mostly fine-grained shale and Siltstone with coarse turbidite intervals deposited in deep water. The Diamond Peak Formation is dominantly sandstone, conglomerate, and siltstone deposited along the margin of the Antler orogenic belt and on the slope descending into the basin axis to the east. The Diamond Peak formation thins to the east, while the interfingering Chainman Shale thickens in that direction. Eastward, both formation pass into shallow-water limestone and siltstone dominated formations such as the Great Blue, Deseret, and Humbug Formations
Late Devonian		Joana Limestone	Transition Sequence: initial foreland basin subsidence and carbonate build-up. Pilot Shale is mostly fine-grained shale and siltstone derived from the Antler orogenic belt and deposited in deep water following early basin subsidence. Joana Limestone represents shallowing of the basin and progradation of the carbonate shelf deposits
		Pilot Shale	
		Guilmette Limestone	Pre-Antler Orogeny: The Guilmette Limestone consists of massive to medium bedded mottled limestone, dolomitic limestone, and subordinate sandstone and sandy limestone intervals. These and correlative strata were deposited on the carbonate platform prior to foreland basin subsidence.

FIGURE 6.5. Selected Devonian and Mississippian formations of east-central Nevada. Vigorous erosion of the uplifted Antler highlands produced a flood of siliciclastic debris into flanking basins.

extended as far west as Utah. This basin, located landward from the Antler belt, is known as the Antler foreland basin (Figure 6.7).

The development of the Antler foreland basin is likely the consequence of the tectonic loading of the lithosphere by the rocks of the Antler allochthon. Once this enormous mass of rock had been stacked up in central Nevada, enough weight was added to the western Laurentian lithosphere to cause it to sink faster than sediment could accumulate in it. Predictably, subsidence rates linked to tectonic loading appear to be greater nearest the Antler belt than farther east. This gave the Antler foreland basin a pronounced asymmetry: it was deepest in the west (closest to the allochthon) and shallowest to the east in central Utah (Figure 6.8). East of the axis of the Antler foreland basin, the foun-

dering of the lithosphere resulted in the west-tilting of the carbonate platform. This tilting lowered the seafloor and increased the depth of the shallow water covering the platform. The rock record of the Antler foreland basin suggests several cycles of subsidence, most likely linked to pulses of uplift in the Antler orogenic belt.

To the far west, on the ocean side of the Antler orogenic belt, the Havallah Basin was a deep ocean sink that also received some of the erosional debris shed from the highlands. This basin did not subside because of tectonic loading, unlike the foreland basin to the east and does not exhibit that basin's asymmetry. Instead, the Havallah Basin was the residual floor of the ocean underlain by thin, dense oceanic lithosphere. Most of the sediments that accumulated in the Havallah Basin were fine-grained

FIGURE 6.6. Conglomerate in the Mississippian Diamond Peak Formation, central Nevada. The large chert pebbles in this conglomerate were likely derived from bedded chert in the Antler allochthon to the west.

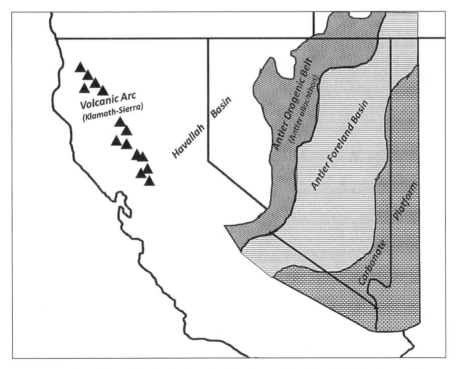

FIGURE 6.7. Location of the Antler foreland basin and the Havallah Basin on either side of the Antler orogenic belt.

siltstone, chert, and argillite. Pillow lavas and volcanic breccia in this deep-water sequence suggest active vents on the seafloor somewhere in or near the deep ocean basin west of the Antler orogenic belt. Sediments accumulated in the Havallah Basin through much of Late Paleozoic time but were displaced and disrupted by later Paleozoic and Mesozoic thrusting and accretionary tectonics. Therefore, there is some lingering uncertainty about the depth, extent, origin, and depositional history of the Havallah basin. To the west of the basin, however, remnants of a volcanic island arc of Devonian-Mississippian age have been recognized in the Klamath Mountains of northern California. This suggests that some of the oceanic lithosphere underlying the Havallah Basin could have been consumed in a magma-generating subduction zone along the extreme western edge of Laurentia (Figures 6.3, 6.7).

Because of the postdepositional disruption of Antler-age sediments in the Havallah Basin, the strata that accumulated in the foreland basin give us the clearest record of sedimentation during and after the Antler orogeny. Complicating matters a bit, though, is the discovery that subsidence of the foreland basin began at a time of rapid and cyclic eustatic sea-level changes. Geologists have identified no fewer than five cycles of rising and falling sea level in the foreland basin rock succession of the eastern Great Basin. Coupled with the periodic subsidence linked to mountain-building, these changes resulted in continuous modifications of seafloor conditions that are reflected in the rock record.

The initial deposits in the Antler foreland basin are most clearly recorded in the coarse sandstone and conglomerate of such rock units as the Tonka and Diamond Peak Formations of central Nevada (Figure 6.8, lower). These sediments consist of fragments of siliceous rocks, such as chert and argillite, eroded from the Antler allochthon and deposited as river gravels and offshore sand and gravel bars along the eastern slope of the Antler highlands. Farther to the east, in the deeper part of the foreland basin known as the foredeep, finer-grained clastic sediments, such as fine sand, silt, and mud, were deposited on the sinking floor of the basin. Such fine-grained materials comprise much of the Late Devonian Pilot Shale and Mississippian Chainman Shale. Turbidite deposits and bedding scours in these formations suggest that some of the sediment arrived in the foredeep via turbidity currents and undersea landslides. Sediment transported by these mechanisms accumulated in deep-sea fans at the mouths of submarine canyons (Figure 6.8).

On the eastern side of the Antler foreland basin, shallow marine limestone and dolomite continued to accumulate on the carbonate platform during the Antler event. However, the numerous Devonian–Mississippian regressions introduced sand and silt from land onto the platform. Also, periodic subsidence in the foredeep tilted the platform to the west, deepening the water at the outer edge. This foundering of the platform is recorded by silty and cherty limestones in rock units such as the Woodman and Deseret Formations (Figure 6.8). Each time the platform sank, the buildup of carbonate sediments eventually restored the shallow marine setting on the platform, and the western edge would prograde offshore (Figure 6.9). Renewed tilting and subsidence, sometimes accompanied by regression or transgression, would begin a new cycle of sedimentation on the carbonate platform. The Late Devonian-Mississippian limestone strata that formed on the carbonate platform commonly contain abundant chert as lenses, nodules, or discontinuous layers (Figure 6.10). The source of the silica (SiO_2) for this chert was probably a combination of volcanic ash from the west, runoff from the siliceous rocks exposed in the Antler highland, wind-blown dust from land, and fluids released from compacting sediments in the deeper parts of the foreland basin.

From its inception in Late Devonian time, about 380 million years ago, the Antler foreland basin collected sediment for some 75 million years before new patterns of oceanic basins were established on the Great Basin seafloor. During this long interval, local subsidence, eustatic changes, paths of sediment dispersal,

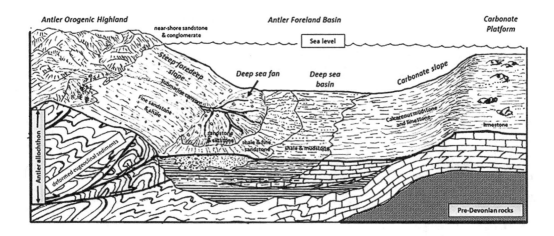

Antler Foreland Basin

West ←———————————— East

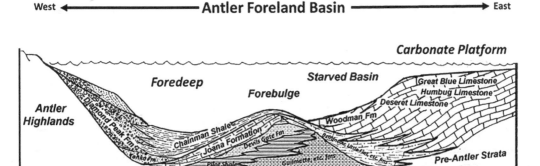

FIGURE 6.8. Depositional setting of the Antler foreland basin and flanking carbonate platform in Mississippian time (*top*) and some representative formations (*bottom*) that developed in it. In addition to those shown, many other formation names have been applied to strata that originated in the Antler foreland basin.

and seafloor flexures continually changed the patterns and rates of sediment accumulation. Flexing in the lithosphere underlying the basin created what geologists have called a migrating forebulge between the foredeep and the carbonate platform on the east (Figure 6.8). The forebulge created a small sag on the sea bottom just offshore (west) of the carbonate platform that was separated from the foredeep by the subtle rise of the seafloor. Almost no siliciclastic sediment shed from the Antler highlands could be transported over the forebulge into the sag, and the water above it was sufficiently deep to limit the formation of carbonate sediments. Deprived of great volumes of sediment from east or west, the sag between the forebulge and the margin of

the carbonate platform is known as a "starved" basin. This means that the rate of sediment accumulation is very low, resulting in a relatively thin sequence of rock layers that record a large increment of geological time. Starved basin deposits consist of black, organic-rich shale, chert, and phosphate minerals in several different forms. Such sediments comprise much of the Deseret and Little Flat Formations of north-central Utah (Figure 6.11) These unusual deposits may have formed when upwelling currents brought nutrient-rich bottom water from the foredeep over the forebulge and into the starved basin setting. This event occurred shortly after a large positive $\delta^{13}C$ excursion is recorded in the Mississippian strata of the

FIGURE 6.9. Cliff-forming limestone of the Joana Formation overlying the Pilot Shale in the Confusion Range of western Utah. The Joana Limestone is typical of the strata deposited on the carbonate platform of the Antler foreland basin. In this location, the edge of the platform prograded westward (*left*) to cover the older deep-water deposits of the underlying Pilot Shale.

FIGURE 6.10. Limestone with irregular nodules and stringers of chert (brown material) in the Devonian Guilmette Formation, Leppy Hills, western Utah.

FIGURE 6.11. Tilted layers of black shale, chert, and phosphate sediment in the Deseret Formation exposed in a trench in the northern Lakeside Mountains, Utah. Such sediments likely accumulated in a starved basin setting within the Antler foreland basin in Mississippian time.

Antler foreland basin. It may indicate that the waters of the foredeep were oxygen depleted, which, in turn, may have limited life on the deep seafloor. Minimal organic absorption of nutrients such as phosphorous could have concentrated these compounds in the foredeep's anoxic bottom water. While there is still considerable debate among geologists on the precise mechanisms by which the phosphate minerals and chert were precipitated, the enrichment of phosphorous is a unique feature of the starved basin sediments.

At times during the Late Devonian-Mississippian interval, the slopes on the forebulge that isolated the starved basin were evidently steep enough to collapse in great undersea landslides. In the south-central Nevada, geologists have identified gigantic blocks consisting mostly of Late Devonian limestone that were shed eastward from the forebulge into fine-grained sediments of the starved basin. These blocks, referred to as olistoliths, were later buried under younger Devonian sediments derived from the Antler orogenic belt. The creation and displacement of these olistoliths in the foreland basin reflects the intense geological activity that accompanied the evolution of the Antler orogenic highland. Earthquakes, turbidity currents, submarine landslides, and cataclysmic slope failures shook the murky depths of the foreland basin for millions of years, while mountain-building was in progress to the west. And yet this geological uproar was only the beginning of the turmoil that would affect conditions on the Great Basin seafloor during the Antler orogeny. Jolts even more dramatic than those related to the Antler orogeny would befall the foreland basin in Late Devonian time. There is abundant evidence in the rock record of the Great Basin that a major catastrophe from the sky was superimposed on the tectonic convulsions of the Late Devonian.

FIGURE 6.12. Outcrop views of the Devonian Guilmette Formation, Hiko Range, Nevada: (A) pre-Alamo Breccia laminated limestone of the lower member; (B) Alamo Breccia Member. Hammerhead in both photos is approximately 13 cm long.

G U I L M E T T E	Upper Member		Limestone, sandstone, and calcareous siltstone containing mud mounds, sponge and coral colonies, brachiopods, and other fossils indicative of normal marine conditions.
	Alamo Breccia Member		Breccia and sandy breccia consisting of large angular blocks of limestone, quartzite, sandstone, and fragmental fossils. Several different units of coarse rubble cut each other and fill scours in underlying strata. Gravel and sand deposits capping the breccia suggest deposition in tsunami surges.
F M	Lower Member		Limestone and calcareous siltstone containing normal marine fauna of sponges, corals, brachiopods, and gastropods. Microbial laminations common.

FIGURE 6.13. Stratigraphic position of the Alamo Breccia Member of the Guilmette Formation, south-central Nevada.

The Alamo Impact Event

In the mid-1990s, geologists studying the Late Devonian strata of south-central Nevada identified an unusual component of the Late Devonian Guilmette Formation in the area around the Pahranagat and Hiko Ranges of southern Nevada. This formation consists mostly of silty limestone deposited along the outer edge of the miogeocline (Figure 6.12A). The limestone contains fossils and structures, including stromatoporoids, brachiopods, corals, and microbial laminations, typical of deposition in normal shallow marine settings. However, in the middle of the formation, an exceptional unit of coarse breccia and conglomerate was discovered that indicates a violent disruption of the calm water over the carbonate platform (Figure 6.12B, 6.13). The coarse rubble contains large fragments and blocks of limestone, quartzite, and sandstone that appear to have been derived from rocks as old as Cambrian. As much as 300 feet thick in some places, the breccia unit could be traced across about 1,500 square miles of the Great Basin, from near Las Vegas to as far north as east-central Nevada. The minimum volume of the breccia is 60 cubic miles, an extraordinary amount of rocky rubble. Because its presence was so consistent over a broad area, this breccia mass was identified as a formal member of the Guilmette Formation in 1997 (Sandberg et al., 1997) and officially named the Alamo Breccia Member after the small settlement of Alamo in Lincoln County, Nevada.

Initially, geologists were uncertain about the origin of the Alamo Breccia. Clearly, some very energetic event shattered the outer edge of the carbonate platform, ripped up a great mass of limestone rubble, and spread the blocky debris across a large area of the Great Basin seafloor. This event might have been a powerful earthquake, a violent landslide on the seafloor, an explosive volcanic eruption, or the impact of an extraterrestrial body. Eventually, detailed study of the Alamo Breccia produced some clues that resolved the mystery of its origin. Quartz grains recovered as insoluble residue when the breccia matrix was dissolved were discovered to possess several intersecting sets of tiny, parallel fractures

that distorted the crystalline structure. Such zones, referred to as shock lamellae, are known to result from shock waves moving through the quartz grains. Shock lamellae occur in quartz grains at locations of underground nuclear detonations and at asteroid impact sites. The discovery of such grains in the Alamo Breccia clearly suggested that sort of violent event.

Also found in the Alamo Breccia were small spherical bodies, up to a quarter inch in size, composed of limestone arranged in concentric layers around a rock fragment nucleus. Similar objects, but of silicate composition, sometimes form in the blast cloud above violent volcanic eruptions. Known as accretionary lapilli, these small grains form in the ejecta cloud of volcanoes when water vapor condenses around an airborne rock fragment in the presence of fine volcanic ash. The very similar objects in the Alamo Breccia were of carbonate composition and likely formed from pulverized limestone dust in the cloud of steam generated by a violent explosion. Yet there are no signs of volcanic activity at the time the Alamo Breccia was formed in the Late Devonian Period. The violent blast responsible for the lapilli must have been triggered by a nonvolcanic mechanism. In addition, small globular silicate and carbonate grains found in the Alamo Breccia appear to have been softened by extreme heat prior to deposition on the seafloor. If volcanic activity is ruled out, a plausible source of heat could have been the energy released by a high-velocity extraterrestrial body striking the shallow seafloor. Intriguingly, sediments from the upper part of the Alamo Breccia contain two to four times more iridium than the background levels. Iridium is an element that is very rare on Earth (only about one to ten parts per trillion in the crust) but much more abundant in some asteroids and meteorites. The increase in iridium in the Alamo Breccia, though relatively small, probably signifies the explosion of an extraterrestrial object that may have occurred on impact or just above the floor of the carbonate platform.

Variations in the grain size, thickness, bedding style, and geometry of the Alamo Breccia deposits suggest that the impact debris was

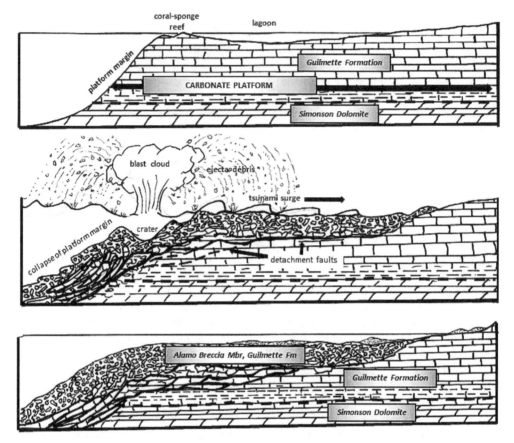

FIGURE 6.14. Origin of the Alamo Breccia Member in the eastern Great Basin region: (*top*) preimpact configuration of the Antler foreland basin in Late Devonian time; (*middle*) after bolide impact, the platform margin collapses into the deeper basin to the west, tsunamis surge across the platform, and subhorizontal detachment faults propagate toward land; (*bottom*) after deposition of the Alamo Breccia, normal marine conditions return, and the upper members of the Guilmette Formation (not shown) accumulate on the carbonate platform.

transported by submarine landslides, powerful tsunamis, and along deep fracture systems (Figure 6.14). Large blocks of limestone and other rocky rubble slid westward from the platform edge into deeper water basins, while huge tsunami surges washed material eastward toward the inner shelf. Layers of sandstone and carbonate conglomerate typically overlie the coarse breccia, indicating that the sea remained turbulent for some time after the initial violence. Overall, the geological evidence supports the conclusion that the Alamo Breccia Member results from the impact of some extraterrestrial body in the shallow seas near the outer edge of the carbonate platform. Assuming such a triggering mechanism, geologists have developed models for the origin and dispersal of the Alamo Breccia (Figure 6.14). While no fragments of the impacting body have yet been identified, it was probably at least 5 km (3 miles) in size to release enough energy to explain all effects recorded in the breccia deposits.

We are still unsure of the nature of this extraterrestrial guest. It could have been an asteroid, a meteoroid, or a comet. Such bodies, when they strike the surface or explode while airborne, are generally known as bolides. If such a violent paroxysm is responsible for the Alamo Breccia, then the catastrophic impact would likely have excavated a crater as much as 50 miles in diameter and a mile or more deep. Where is this crater, and where exactly did the celestial object wallop the Great Basin seafloor? Unfortunately, no crater and impact point can be precisely identified. Later geologic events, and perhaps some that accompanied the actual cataclysm, have made it almost impossible to identify ground zero with certainty. It is possible that the outer edge of the carbonate platform collapsed upon impact, or soon after. If so, the crater would have spread out laterally, and a portion might have disintegrated inward and/or down the nearby continental slope. Also, even if a well-formed crater were excavated during the Late Devonian, later geological events in the Great Basin would likely have obscured or buried it. The thrust faulting that characterized the nearby Antler orogeny continued well into the Mississippian subperiod, and the site of the impact may have been buried under sheets of rock transported from the west.

Much later, during the Mesozoic era, thrust faults again became active in central Nevada, and several sheets of rocks were moved eastward toward Utah, piling up about where the impact should have occurred. Then, during the Cenozoic Era, the tensional forces that shaped the modern mountains of the Great Basin ripped the crust, and the buried crater would have been dismembered, with pieces lifted or buried in fault-blocks of the stretched and broken crust. Nonetheless, breccia-filled fractures, slope blocks, and ring faults associated with the Alamo Breccia in the area near the Timpahute Range of southern Nevada are thought to indicate a portion of the crater rim (Figure 6.15). These features do not represent the crater rim itself but only suggest sleep slopes and shattered bedrock related to it. In addition, detailed studies of the thicknesses and nature of sediments deposited directly above the Alamo breccia allow scientists to recognize the general shape and size of the bowl-like depression in which the postimpact debris accumulated. Estimates of the crater size, prior to its burial and tectonic mutilation, range from about 25 to 100 miles across.

The effects of the Alamo event can be recognized in Devonian rocks far beyond the likely impact site. Slumps and dislodged masses of sediment and rock have been identified in portions of the Guilmette Formation in White Pine County, some 130 miles to the north. In southern Nevada, exposures of the Devonian Sultan Formation at Frenchman Mountain, 100 miles southeast of the impact site, include an interval of breccia, deformed strata, and rocky rubble within the sequence of normal shallow marine dolomite and sandstone. These deposits probably represent material destabilized because of powerful tremors associated with the impact. The strong vibrations from the impact-generated earthquakes appear to have shattered the rocks and mobilized masses of soft sediment into fluidized submarine avalanches. Such widespread evidence of sediment disturbance suggests that the event created turbulence and turmoil that swept great distances across the Great Basin seafloor.

Life after the Alamo Impact

For the organisms living in the Antler foreland basin, the Alamo impact was certainly catastrophic. Recall, also, that the Late Devonian extinction discussed in Chapter 5 followed soon after the calamity from space, according to current age estimates for both events. Furthermore, some scientists have noted that several other impact sites around the world appear to be close in age to the Alamo event (Barash, 2016, listed in Chapter 5 references). Could this suggest some sort of meteor or asteroid storm triggered the early phases of the Late Devonian extinctions? Maybe. In any event, the end of the Devonian Period was a difficult time for marine organisms of the Great Basin seas.

Still, fossils from postimpact Late Devonian and Mississippian strata in the Great Basin suggest a rapid recovery of the marine biota following the ecological chaos of the Late Devonian. Almost immediately (geologically speaking, anyway) after the impact-related deposits settled across the shattered basement of the Great Basin seafloor, deposition of normal carbonate sediments resumed and continued into the

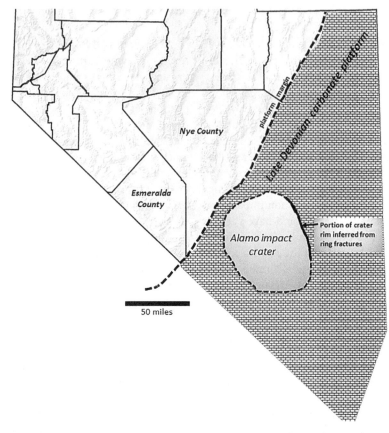

FIGURE 6.15. Approximate location of the Alamo impact inferred from stratigraphic trends, ring fractures, and near-rim deposits.

FIGURE 6.16. Conger Mountain in the Confusion Range of western Utah. The lower slopes are eroded outcrops of the highly fossiliferous Chainman Shale of Mississippian age. The overlying cliffs are exposures of the Ely Limestone of Pennsylvanian age.

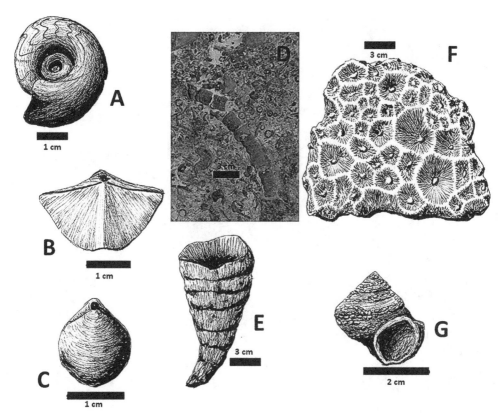

FIGURE 6.17. Common Mississippian marine invertebrates of the Great Basin: (A) *Cravenoceras*, a goniatite cephalopod genus; (B) *Spirifer centronatus*, a brachiopod; (C) *Composita* sp., a brachiopod; (D) disarticulated crinoid stem segments and ossicles, Chainman Shale, Nevada; (E) a compound rugose coral; (F) *Glabrocingulum* sp., a common gastropod genus.

Mississippian subperiod. Fossils are exceptionally abundant in these postimpact sediments, including organisms from a variety of environments within the Antler foreland basin such as the foredeep, backbulge basin, and carbonate platform. Among the many Great Basin formations that produce abundant Mississippian fossils are the Chainman Shale of western Utah, the Joana-Lodgepole Limestone formations of west-central Utah, the Great Blue Limestone of central Utah, and the Monte Cristo Limestone of Nevada. In some places, such as Conger Mountain of western Utah, Mississippian fossils occur in spectacular concentrations (Figure 6.16).

Corals, sponges, brachiopods, gastropods, cephalopods, conodonts, and crinoids were all common in the shallow marine communities that repopulated the Antler foreland basin

after the Alamo impact (Figure 6.17). Corals are prominent as reef-building organisms, particularly along the outer banks of the carbonate platform. Both solitary and compound corals (Figures 6.17, 6.18) became the main reef-building invertebrates in the Mississippian seas after the severe reduction of stromatoporoids and sponges during the Late Devonian extinctions. Coiled cephalopods were also widespread in the Mississippian seas of the foreland basin. Overall, the Mississippian invertebrate assemblages are very similar to the earlier Devonian faunas, even though many new species evolved in the various lineages to replace those lost to extinction at the end of the Devonian Period.

Coiled cephalopods of the molluscan subclass Ammonoidea were so abundant and widespread in the Mississippian seas of the foreland

FIGURE 6.18. Jumbled remains of solitary corals in the Mississippian Great Blue Limestone, Lakeside Mountains, western Utah.

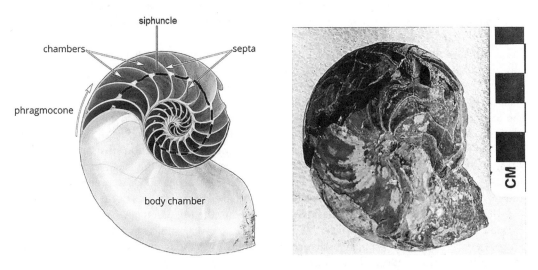

FIGURE 6.19. Generalized coiled cephalopod shell structure (*left*) and *Goniatites deceptus*, an ammonoid cephalopod from the Mississippian Chainman Shale of western Utah (*right*). Diagram from the Paleontological Research Institution, Ithaca, NY; fossil specimen from the Natural History Museum of Utah (IP 4136).

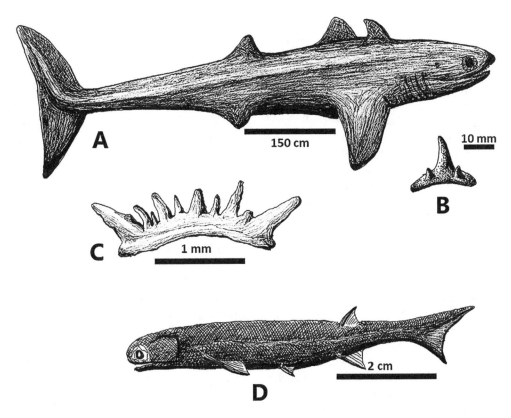

FIGURE 6.20. Selected Mississippian fish from the Great Basin: (A) reconstruction of a cladodont shark similar to those that bore the multicusped teeth in B; (B) a typical, isolated cladodont shark tooth from the Manning Canyon Shale; (C) the comb-like teeth of *Diademodus* from the Pilot Shale of western Utah; (D) a reconstruction of *Utahacanthus*, a minnow-sized acanthodian fish.

basin that they deserve special mention. These swimming predators possessed internal partitions (septa) that divided the spiral shell into numerous chambers, not unlike the shell of the modern nautilus (Figure 6.19). The living animal resided in the largest and outermost chamber, with the inner chambers being occupied during earlier life stages. The chambers were connected with a fleshy tube called the siphuncle that served to regulate the fluid and gas pressure in the vacated chambers. This gave the ammonoids buoyancy control as they swam through the water in search of their prey. Scores of ammonoid species have been identified from the Mississippian strata of the Great Basin, especially in the Chainman Shale, Scotty Wash Formation, and lower part of the Ely Limestone. Because they were efficient swimmers

and not restricted to specific bottom conditions, ammonoid fossils are widespread in the Great Basin and occur in sedimentary rocks from a variety of depositional environments. Such considerations make them valuable for correlation and subdivision of the Mississippian rock record from Utah to California.

There were also several types of fish cruising the Mississippian ocean across the Great Basin seafloor, including the reefs, lagoons, and inlets adjacent to the Antler highland and the swamps of central Utah (Figure 6.20). Though fish fossils are relatively uncommon in Mississippian strata, enough have been found that it is possible to document the presence of several types descended from Devonian ancestors. Isolated teeth from several different kinds of sharks have been found in the Manning Canyon Shale and Great Blue Formation. Sharks, like other chondrichthyan fish, had skeletons composed primarily of soft cartilage, which is prone to destruction and decomposition after death. Teeth, on the other hand, consist of durable organo-phosphatic material and are much more likely to be preserved as fossils. Also, all sharks had scores of teeth that were continuously shed and replaced during life. An individual shark could have produced thousands of highly pre-servable teeth over its lifetime but only one frag-ile skeleton. It is not surprising that nearly all the Mississippian sharks identified from the Grand Canyon to Montana are known primarily based on teeth. Nonetheless, given the amazing diver-sity of shark teeth recovered from Mississippian strata of western North America, it appears that many kinds of primitive sharks flourished in the Antler foreland basin.

Some sharks of the Antler foreland basin seas belong to a group known as the cladodonts. These sharks possessed teeth with one large spike surrounded by a variable number (usually two or four) of smaller denticles (Figure 6.20 A–C). Such teeth are quite different from the large, sharp-edged slicing teeth of modern sharks. Cladodont teeth appear well suited to capture actively swimming prey, and to prevent their escape. Most cladodont sharks were about 2–3 feet long and probably fed on smaller fish and swimming cephalopods. Shark teeth of this type have been found in rocks deposited in open marine and reef environments as well as in near-shore mudstone and shale, which suggests that these ancient sharks were tolerant of brackish marsh water as well as normal seawater. Accom-panying the sharks were several acanthodian fish, also descended from Devonian ancestors. The best known of these is *Utahacanthus* (Fig-ure 6.20D), a small, minnow-sized fish. Acti-nopterygians (ray-finned fish) were also quite varied in the Manning Canyon shale of central Utah, from which at least three different genera are known. They were generally less than a foot long, had tiny, sharp teeth, and were likely active swimmers. They may have fed on a variety of prey organisms including (see below) insects, amphibians, aquatic larvae, and small fishes in the near-shore waters where they lived. Though

these fish fossils are not the most abundant ele-ments in the Mississippian fossil fauna, they do suggest that many kinds of fish were well adapted to a variety of marine and marginal marine habitats. Future discoveries will, no doubt, reveal a more complete picture of the diversity of fish that swam the seas of the Antler foreland basin.

Mississippian Life on Land

Because the deformation and uplift of the Ant-ler orogenic belt remained active through much of Mississippian time, land areas in central Nevada continued to rise. Vigorous erosion of the mountainous Antler orogenic belt produced a great volume of gravel, sand, and silt that was transported to the foreland basin. Gradually, the erosional refuse that collected near land constructed deltas and coastal plains beyond which many bays, inlets, and lagoons connected with the open ocean. On the eastern side of the foreland basin, the shoreline was situated in east-central Utah by Late Mississippian time. A broad zone of fresh-water swamps and brackish marshes separated the land in what is now the Rocky Mountain region and the open seas of the foreland basin farther west. This low, boggy zone between land and sea resembled the mod-ern Everglades in south Florida. Thousands of feet of organic-rich shale, silt, sand, and mud accumulated in the swamps and coastal plains along the eastern and western sides of the Antler foreland basin. These sediments comprise the finer-grained portions of the Diamond Peak Formation of Nevada and the Manning Can-yon Shale and Doughnut Formations of west-ern Utah. Plant and animal fossils from these formations give us a remarkable view of the ear-liest rich terrestrial flora and fauna in the Great Basin region.

Mississippian plant fossils document the presence of coastal forests resembling modern mangrove and cypress jungles along both sides of the foreland basin. However, these ancient forests were composed of entirely different types of trees and shrubs. The jungle understory was comprised of primitive plants such as ferns, mosses, and horsetails. The most dominant large plants were the primitive trees known as

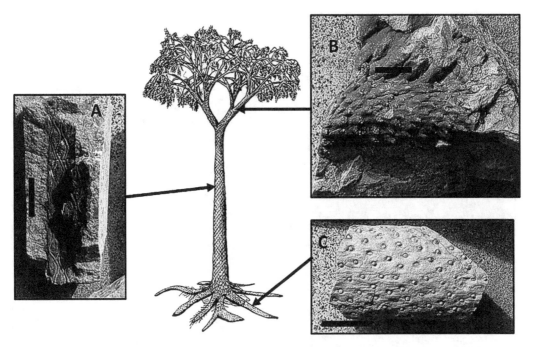

FIGURE 6.21. *Lepidodendron* fossils from the Diamond Peak Formation, central Nevada. Middle sketch depicts the mature tree standing over 100 feet tall.

lycopods, or scale trees. The largest and most distinctive extinct scale-tree is *Lepidodendron*, which grew in thick stands in wet swampy conditions (Figure 6.21). The scale trees are so named because their trunks had scaly leaves arranged in vertical rows of spirals. When the leaves were detached, diamond shaped scars were left on the trunk's outer covering. The lycopods were crowned by branches that bore long narrow leaves, some of which ended in cone-like structures. The branched roots were shallow and had small rootlets arranged in spiral fashion that collected water and nutrients. The largest *Lepidodendron* trees stood some 150 feet tall, with trunks several feet in diameter near the base. The shallow root system consisted of numerous branches bearing smaller rootlets. Root segments are commonly preserved as fossils known as *Stigmaria*, easily identified by the distinctive spiral arrangement of rootlet scars (Figure 6.21). Because leaves were present on the trunks as well as the branches of *Lepidodendron*, the entire plant was green. The Lycopod-horsetail-fern forests were widespread along the coastal lowlands of the Antler highland, and

fossils of these trees have been found in Mississippian and Pennsylvanian rocks from Death Valley to northern Utah. Very similar forests grew along the eastern margin of the Antler foreland basin, where their remains are strikingly abundant in the black, brown, and yellowish shales of the Manning Canyon Formation.

The lush coastal forests of the Antler foreland basin also supported a variety of terrestrial animals. Though their fossils are not nearly as abundant as those of the plants in Mississippian strata of the Great Basin, scientists have found ample evidence that the ancient forests and woodsy lagoons were swarming with animals. Fossils of insects like modern dragonflies have been found in the Manning Canyon Shale, suggesting that flying bugs buzzed through the swamps and jungles where the muddy sediments accumulated. Also, the remains of a small salamander-like amphibian known as *Utaherpeton* have been discovered in the upper part of the Manning Canyon Shale (Figure 6.22). This tiny crawler was only about 1.5 inches long and resembled a salamander. It had very simple, peg-like teeth along its short jaws.

FIGURE 6.22. A coastal forest and swampy lagoon along the margins of the Antler highland in Mississippian time. Inset: reconstruction of the skeleton (*top*) and life appearance (*bottom*) of *Utaherpeton franklini* from the Manning Canyon Shale of western Utah. Scale bar = 1 cm.

Utaherpeton seems poorly suited to consume the coarse vegetation that was dominant in the swampy forests where it lived. Perhaps it fed on insect eggs or larvae, small fish or amphibians, soft fern leaves, organic detritus, or some combination of these. In addition to *Utaherpeton*, a partial skeleton of another small, amphibian-like creature is known from the Diamond Peak Formation in central Nevada. Given the genus name *Antlerpeton*, the skeletal material was preserved as impressions in sandstone, including much of the backbone, pelvis, and hind limbs. *Antlerpeton* was probably at least 6 inches long in life and had hind limbs and a robust pelvis that seem suited for crawling on land. However, this creature was very primitive in other skeletal aspects and may be transitional between sarcopterygian fish and true amphibians. The taxonomic affinity of *Antlerpeton* is still uncer-

tain, and more complete specimens are needed to resolve questions about its evolutionary status. For now, paleontologists refer to *Antlerpeton* as a "basal tetrapod," meaning that it is one of the earliest vertebrates to make the transition from water to land. *Antlerpeton* was probably not fully terrestrial and stayed close to the water around swamps and lagoons.

Though fossils of terrestrial vertebrates are rare in the Mississippian strata deposited in the foreland basin, these specimens clearly demonstrate that at least some of the descendants of the Devonian fish had moved far along the evolutionary path leading to land life by Late Mississippian time. This transition was a global phenomenon, but it is intriguing to ponder how quickly life moved ashore once land emerged from the Great Basin seafloor.

7

The Antler Overlap Sequence

The final two phases of the Paleozoic Era are the Pennsylvanian subperiod of the Carboniferous Period (323–299 million years ago) and the Permian Period (299–252 million years ago). This 75-million-year interval was a time of great changes in global geography, climate, and life. The vast supercontinent of Gondwana continued its crawl to the north and, in the later Paleozoic, began a slow-motion collision with the southeastern margin of Laurentia (or Laurussia). The collision resulted in the final pulse of mountain-building, a stage of deformation and uplift known as the Alleghenian orogeny in eastern North America and the last in a sequence of tectonic upheavals that constructed the ancestral Appalachian Mountain system. Unlike the modern Appalachians, however, this ancient massif was an inland mountain chain, stretched from Newfoundland to west Texas and rising to heights that would rival the modern Himalaya Mountains (Figure 7.1). For millions of years after their emergence, erosion of these lofty peaks provided sediment that ultimately spread over most of what is now North America, including the Great Basin and Rocky Mountains far to the west. The collision of Laurentia with Gondwana was also the principal tectonic event that, by the end of the Paleozoic Era, resulted in the amalgamation of Pangaea, the largest and most recent supercontinent in earth history.

The powerful forces generated by the collision of Gondwana and Laurentia also penetrated well into the interior of both continents. In what is now the Rocky Mountain region, the Laurentian crust buckled and tore along deep faults where several prominent mountain blocks lifted. Collectively, these uplifts are known as the Ancestral Rocky Mountains, which consisted of numerous uplands of varying orientation and size (Figure 7.2). The blocks of rock that lifted nearest the Great Basin region included the northwest-trending Uncompahgre uplift of eastern Utah and Colorado and the broad Emery uplift of central Utah and northern Arizona (Figure 7.2). Between the lofty ramparts of the Ancestral Rocky Mountains, fault-bounded basins subsided and filled with sediment eroded from the adjacent highlands. While none of the uplifts associated with the Ancestral Rocky Mountain orogeny rose in the Great Basin region, several Pennsylvanian and Permian basins, like those of Colorado and Utah, developed on the floor of the Antler foreland basin. These localized, Late Paleozoic basins, such as the Oquirrh, Bird Spring, and Ely Basins were ultimately filled by thick sequences of limestone, siltstone, and shale. On the western side of the eroded Antler highland, the Havallah Basin continued to subside, accumulating great volumes of sediment during the Late Paleozoic Era. Other parts of the Great Basin seafloor remained submerged under shallow water, continuing the long pattern of deposition that characterized the carbonate platform of the Antler foreland basin.

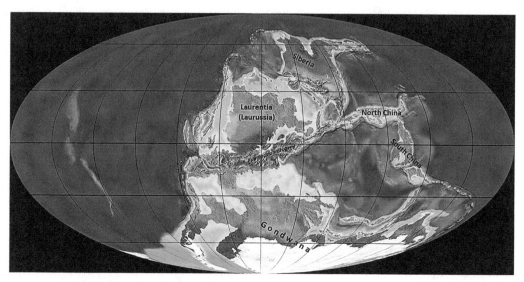

FIGURE 7.1. Global geography during the Pennsylvanian subperiod. Paleogeographic base map prepared by R. Blakey, Colorado Plateau Geosystems Inc.

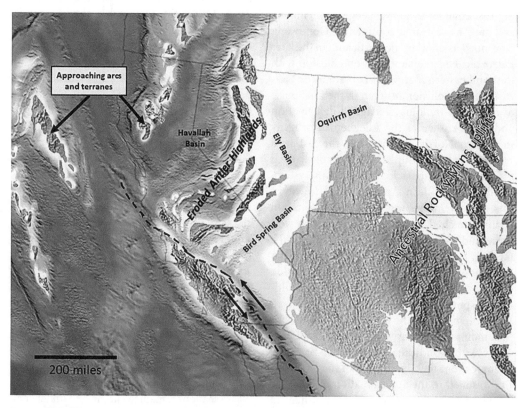

FIGURE 7.2. Western Laurentia during Middle Pennsylvanian time. Paleogeographic base map prepared by R. Blakey, Colorado Plateau Geosystems Inc.

While the eastern margin of Laurentia was colliding with Gondwana, plate tectonic interactions along the western edge of the continent appear to have been more complicated. In general, the convergent plate interactions that were initiated during the Antler orogeny continued into the later Paleozoic Era along much of the continental margin of the modern Great Basin region. Though the main phase of mountain-building had passed in the Antler orogenic belt, more localized thrust faulting and deformation continued well into the Pennsylvanian subperiod—and perhaps even later. Areas lifted by post-Antler deformation were vigorously eroded, and coarse, clastic sediment accumulated around the rising land. However, there is also evidence of extensional forces immediately west of the subdued Antler highlands, where the Havallah Basin sank as the crust beneath it was evidently stretched. The deep-water sediments (mostly siltstone, shale, and chert) that accumulated in this basin include volcanic materials, which suggests either the presence of a nearby oceanic volcanic island arc or the possible undersea eruption of lava within the basin itself. Because rocks of the Havallah sequence were later thrust eastward, the exact size and tectonic setting of the basin remains unclear.

To make matters even more complex, recent studies have identified rocks in Sonora, Mexico, that appear to have been transported nearly 2,000 miles southwest from the southern Great Basin region. This observation may be evidence that lateral (side-by-side, or "transform") plate tectonic movements sliced away a fragment of the southwestern edge of Laurentia—including the southern Great Basin region—and displaced it far southward. Thus, all three of the fundamental stresses—compression, tension, and shear—appear to have affected the Great Basin region during the final phases of the Paleozoic Era. Given this complexity, geologists have yet to work out the precise mechanisms for the development of all the highlands and basins across the Great Basin during the Antler overlap sequence. The task is made very difficult by the later disruption of these rocks and geological structures in the Mesozoic Era, the subject

of Chapter 8. Yet we can be confident that the plate tectonic setting of the Great Basin seafloor in the Late Paleozoic was anything but tranquil or consistent.

In the portions of the Antler highland that were not affected by strong uplift, erosion in Late Paleozoic time lowered the summits, reduced the extent of land area, and produced sediment that accumulated in both the foreland and Havallah basins. By Middle Pennsylvanian time, so much sediment had accumulated in the foreland basin that the foredeep portion was far less prominent than it had been in the Devonian and Mississippian. The progressive filling of the foredeep, coupled with the outward growth of the carbonate platform, restricted deep-water conditions to the smaller basins stretching from Idaho to southern Nevada. As we will soon see, these changes greatly influenced the patterns of sediment accumulation during Late Paleozoic time.

The Late Paleozoic Ice Age and Sea-Level Oscillations

Late in the Mississippian subperiod, global sea level fell to end the Kaskaskia sequence of inundation across most of Laurentia (Figure 4.2). This decline exposed vast areas of the continental seacoasts, resulting in erosion that in turn produced a widespread unconformity separating the Late Mississippian and Early Pennsylvanian strata. The global regression of the seas reflects the combined effects of mountain-building accompanying the formation of Pangaea, Late Paleozoic disruptions of the terrestrial biosphere, and the onset of a new global ice age.

We have already seen that lush forests of primitive trees and shrubs moved onto land in the Great Basin in Mississippian time. This important step in plant evolution and adaptation occurred across the globe at about the same time. During the Pennsylvanian subperiod, carbon-rich organic matter from these forests accumulated in swampy environments on a global scale, forming vast deposits of coal worldwide. So rich in carbon are the sediments of Mississippian and Pennsylvanian age that these two subperiods comprise the Carboniferous

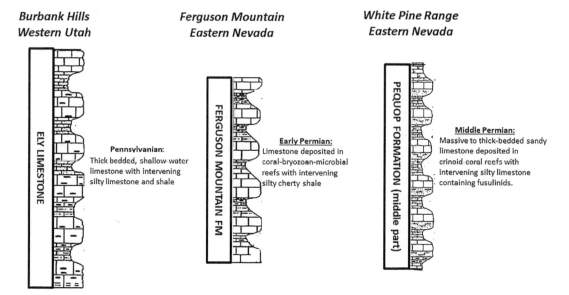

Burbank Hills
Western Utah

Ferguson Mountain
Eastern Nevada

White Pine Range
Eastern Nevada

ELY LIMESTONE

FERGUSON MOUNTAIN FM

PEQUOP FORMATION (middle part)

Pennsylvanian:
Thick bedded, shallow-water limestone with intervening silty limestone and shale

Early Permian:
Limestone deposited in coral-bryozoan-microbial reefs with intervening silty cherty shale

Middle Permian:
Massive to thick-bedded sandy limestone deposited in crinoid-coral reefs with intervening silty limestone containing fusulinids.

FIGURE 7.3. Cyclic layering in selected Pennsylvanian and Permian strata of the Great Basin. Note the rhythmic alternation of layers of limestone, shale, and silty limestone in these examples. Such patterns reflect regular changes in bottom conditions linked to oscillations of water depth and sea level.

Period, a name recognized worldwide. The burial of organic carbon as coal sequestered enormous amounts of carbon from the carbon cycle, eventually lowering levels of carbon dioxide in the atmosphere. Meanwhile, the assembly of Pangaea was lifting major mountain systems wherever landmasses were colliding, exposing submerged areas and invigorating the erosion of bedrock across Pangaea. The chemical reactions that drove the accelerated weathering process drew atmospheric carbon dioxide down even further.

The decline of atmospheric carbon dioxide initiated a strong cooling trend starting about 340 million years ago (Middle Mississippian time) that would culminate in widespread glacial conditions through the remainder of the Paleozoic. However, the pattern of glaciation during the Late Paleozoic ice age was not simple. It began as a series of brief cold intervals that spawned small glaciers in parts of Gondwana. Each of these intervals was separated by warmer interludes lasting perhaps a few million years. By the end of the Mississippian subperiod, ice masses in Gondwana had grown large enough

to cover extensive areas, causing a global decline in sea level. Throughout the subsequent Pennsylvanian subperiod, ice masses expanded and contracted in concert with erratic shifts in the global climate. Sea level fell with each cycle of glaciation and rose during the warm interglacial periods. The rhythmic rise and fall is recorded in sedimentary rocks worldwide, including the great Basin region, as very cyclic patterns of sediment accumulation (Figure 7.3).

Some 300 million years ago, near the transition from Pennsylvanian to Permian time, the Late Paleozoic ice ages reached their maximum. This protracted decline in sea level characterizes the latter portion of the Absaroka sequence (Figure 4.2). Glacial ice covered large areas in the higher latitudes of southern and northern Pangaea. Most of western North America, then positioned about 25 degrees north of the equator, was not directly affected by the glaciation, although glaciers may have developed in the higher ramparts of the Ancestral Rocky Mountains such as the Uncompahgre uplift. Nonetheless, the oscillations of sea level had a pronounced influence on the sediments that

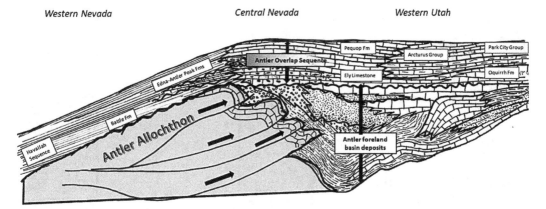

Western Nevada   Central Nevada   Western Utah

FIGURE 7.4. Simplified diagram of the Pennsylvanian–Permian Antler overlap sequence with representative formation names. The heavy undulating lines represent regional unconformities and gaps in the rock record resulting from periods of nondeposition and/or erosion.

accumulated across the Great Basin seafloor in Late Paleozoic time. After Middle Permian time, warming trends caused rapid deterioration of the global ice sheets to the end of the period. Sea level rose at the dawn of the Mesozoic Era, reversing the long, erratic decline that characterized much of the Late Paleozoic Era.

Rocks of the
Antler Overlap Sequence

In the mid-to-late 1900s, geologists working in the Great Basin noted that the strata displaced eastward during the Antler orogeny were buried in many places under younger Paleozoic deposits of both marine and terrestrial origin. Typically, the contact between the rocks of the Antler allochthon and the overlying sediments was marked by an irregular unconformity that represented an ancient erosion surface. This suggested that after their initial uplift millions of years earlier, the Antler highlands had been lowered enough by erosion in Pennsylvanian time that sediment once again accumulated over parts of the lower terrain. Because the Pennsylvanian and Permian sediments buried the older deformed rocks of the ancient highland, the series of younger strata became known as the Antler overlap sequence (Figure 7.4). Many different formations comprise the Antler overlap sequence, and the sediments accumulated in a variety of depositional environments, ranging

from alluvial fans and sandy riverbeds to coastal deltas and swamps, to shallow open seafloor, to deeper ocean basins.

Recent studies of the rocks comprising the Antler overlap sequence have revealed complexities that evince a much more complicated geological history than originally thought. Aside from the underlying unconformity, the entire overlap sequence has numerous other internal unconformities that suggest faulting and uplift continued, at least locally, far beyond the traditional "end" of the Antler orogeny in Mississippian time. Some of the unconformities are also likely related to the erratic eustatic sea-level fluctuations associated with the Late Paleozoic ice ages. The unconformities are recognized in the rock sequence as an angular discordance between sets of strata (see Supplemental Field Trip #1), as irregular surfaces of erosion that cut out some layers, as reddish-brown ancient soils separating unweathered strata above and below, and as zones of missing fossils in a known pattern of occurrence. Each of the unconformities signifies erosion linked either to uplift generated by plate tectonic forces, sea-level decline, or some combination of both factors. Geologists have identified no fewer than 10 major unconformities between and within the various formations comprising the Antler overlap sequence. There is still considerable uncertainty about precise age and origin of these unconformities.

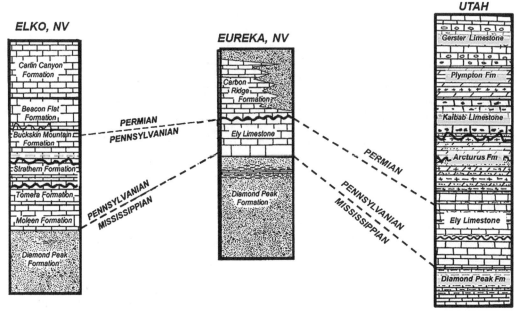

FIGURE 7.5. Unconformities (heavy irregular lines) in the Antler overlap sequence indicate ongoing tectonic activity, uplift, and erosion in the Great Basin region during the latter periods of the Paleozoic Era.

FIGURE 7.6. Outcrop (*background*) and close-up (*inset*) views of tilted conglomerate layers in the Permian Garden Valley Formation, Eureka County, Nevada.

FIGURE 7.7. (A) Outcrop of the Permian Gerster Formation in the Confusion Range, western Utah. Inset at the upper right is a close-up of the limestone with circular crinoid ossicles and other fragmentary fossils; (B) tan chert nodules in the Pennsylvanian Ely Limestone, south Egan Range, Nevada. The chert masses are less susceptible to weathering and typically protrude from the limestone host rock.

It is clear, however, that the Pennsylvanian and Permian on the Great Basin seafloor were not times of uniform submersion and geological quiescence.

An additional indication that geologic upheavals continued into the Late Paleozoic in the central Great Basin lies in several prominent horizons of coarse conglomerate in the overlap sequence. The conglomerate intervals typically grade laterally into river-deposited sandstones and, where they can be traced far enough, into deltaic sandstone and mudstone. The coarse, conglomerate sediments occur in many different horizons within the overlap sequence of north-central Nevada, including the Tonka, Tomera, Strathern, Battle, and Garden Valley Formations (Figure 7.6). These materials represent rocky rubble shed from rising uplifts and deposited by streams in alluvial fans, river channels, and sandbars. All the intervals of coarse sediment in the overlap sequence are restricted to areas within or near the remnant Antler orogenic belt. Many of the conglomerate deposits

are associated with the regional unconformities, suggesting a genetic association between the two indications of uplift.

Beyond the highlands, most of the Great Basin remained submerged during Pennsylvanian and Permian times. Where the water was shallow and warm, limestone and dolomite sediments are dominant and widespread in numerous rock units in the overlap sequence. Examples of such shallow-water carbonate strata include the Callville Limestone, Antler Peak Limestone, Park City Formation, Plympton Formation, Gerster Formation, and Kaibab Limestone. Shallow marine invertebrate fossils typical of the Paleozoic fauna are commonly found in these carbonate strata (Figure 7.7A). Other formations of marine origin, such as the Weber Quartzite, Ely Formation, Oquirrh Formation, and Bird Spring Formation consist of mixed rock types in which limestone layers alternate with intervals of silty limestone, shale, and sandstone. Chert, in the form of brown or black nodules and oblong bodies, is notably abundant in many of the limestone and shale layers of the overlap sequence formations, especially those of Permian age (Figure 7.7B). The formation of chert from silica (SiO_2) dissolved in seawater is usually very limited in warm, shallow seas where carbonate sediments form most readily. So, how did so much chert originate in the Late Paleozoic seas of the Great Basin?

After sponges die, the tiny siliceous spicules they use to support the soft tissues can nucleate microcrystalline quartz (chert!) after they accumulate on the seafloor. While sponges were certainly present on the Great Basin seafloor during Pennsylvanian and Permian times, they do not seem to be sufficiently abundant to be the source of all the chert preserved in rocks of that age. However, there may have been another source of siliceous sediment: the vast interior deserts of western Pangaea. By Permian time, western Laurentia, embedded into Pangaea, was located between 10°N and 30°N latitude. The Late Paleozoic glaciers were most prevalent at higher latitudes, but the subtropical interior of Pangaea was notably arid, with vast dune fields covering extensive tracts of land. The dune-deposited sandstones of Pennsylvanian-Permian age in the Colorado Plateau and Rocky Mountain regions are the sedimentary record of these ancient deserts. Persistent trade winds over the arid interior transported great clouds of dust from the deserts westward to the offshore oceans in the Great Basin, much like the dust storms that periodically drift west today from the deserts of North Africa out across the Atlantic Ocean. Such wind-blown dust consists primarily of tiny grains of quartz and other silicate minerals. In the Pennsylvanian-Permian, it could have provided much of the parent material for the chert nodules in the Antler overlap sequence limestones.

Another curious aspect of the rocks on the Antler overlap sequence is the abundance of calcium phosphate minerals in the carbonate and siliceous sediment that accumulated in the northwestern portion of the Great Basin during the Permian Period. Geologists have applied the name Phosphoria Formation to a sequence of Permian limestone, siltstone, shale,

mudstone, and dolomite that contains such high concentrations of phosphate minerals that some intervals are referred to as phosphorite deposits. Phosphorite is any sedimentary rock in which phosphate minerals, usually fluorapatite [$Ca_5(PO_4)_3F$], are sufficiently abundant that they can be mined profitably as a source for phosphorous. The phosphate minerals in the Phosphoria Formation are present as small, spherical concretions (known as ooids), pellets, and cementing minerals in limestone, chert, shale, or mudstone. In the northeastern Great Basin, the host sediments were deposited in shallow water near the edge of a deeper basin to the southwest. Most geologists attribute the unusual concentration of phosphate minerals to upwelling currents that brought deep, nutrient-rich bottom water up a submarine slope toward a shallow carbonate platform. In turn, these currents were likely generated by offshore winds blowing from land located in what is now Montana (Figure 7.8). In the warmer shallow water, the nutrients introduced by the upwelling currents would have stimulated a phytoplankton bloom in the seas at the edge of the platform. The residue of planktonic organisms fell to the shallow seafloor, where oxygen was consumed as the organic matter decomposed. The phosphate minerals were likely precipitated, perhaps with the aid of organisms, in the shallow water or in the anoxic, organic-rich mud on the seafloor. The Phosphoria Formation has been the main target of more than 50 years of mining activities in the great Western Phosphate Field that stretches from Nevada to Montana (Figure 7.8). The phosphorite strata may also be the source rock for much of the petroleum underlying northern Utah and western Wyoming.

In addition to the phosphate minerals and petroleum, the Phosphoria Formation sediments are also rich in uranium and vanadium. All these components would have been abundant in the water rising from the deeper basins and are compatible with the concept of upwelling currents as a primary factor in their origin. While a complete understanding of the Permian phosphorite deposits is still under study, it seems that the juxtaposition of shallow car-

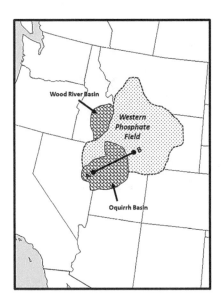
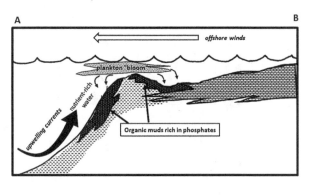

FIGURE 7.8. Map of the Western Phosphate Field (*left*) and a generalized model for the origin of phosphorite deposits in the Permian Phosphoria Formation (*right*).

bonate ramps adjacent to the deeper basins was critical in the formation of these unusual sediments. The rhythmic pattern of deposition seen in other Late Paleozoic rock sequences is also evident in the phosphorite strata of the Phosphoria Formation, suggesting that bottom conditions changed regularly with the Permian eustatic fluctuations.

Late Paleozoic Basins of the Great Basin

While most of the Great Basin seafloor was submerged beneath only a few hundred feet of water during the Pennsylvanian and Permian, several areas subsided rapidly to form deep basins where great thicknesses of sediment accumulated. The most prominent of these sinks was the Oquirrh Basin centered in northwestern Utah (Figure 7.2). This basin began to subside in the Mississippian subperiod but sank so rapidly in Pennsylvanian time that more than 20,000 feet of sand, silt, and carbonate mud accumulated during that period alone and persisted, in a more restricted and subdued form, into the Early Permian Period. The Oquirrh Group of western and northern Utah consists of several formations and constitutes the primary rock record

of the Oquirrh Basin. This thick succession of strata includes alternating intervals of laminated mudstone, calcareous siltstone, bioclastic limestone, and fine sandstone. The sediments of the Oquirrh Group accumulated in both shallow and deep water, a consequence of the interplay between sea-level fluctuations and rates of basin subsidence. Sandy deposits accumulated when water depths were minimal, and granular materials reached the basin from deltas in central Utah, while finer-grained muddy sediments were deposited under deeper water conditions (Figure 7.9). The carbonate mud represented by the limestone intervals was likely transported by bottom currents sweeping over the shallow-water rim surrounding the abyss.

In east-central Nevada, the Ely basin received more than 6,000 feet of sediment during the Pennsylvanian subperiod. Much of the sediment consisted of carbonate mud deposited under shallow-water conditions, indicating that the subsidence was less extreme than in the Oquirrh Basin to the northeast. The Ely Limestone or Group (some geologists have grouped it with other rock units of Permian age) consists primarily of tan to medium-gray limestone intervals alternating with calcareous

FIGURE 7.9. Calcareous mudstone in the Oquirrh Group of the Oquirrh Mountains of northwest Utah. The dark markings are trace fossils recording the movement of organisms through the mud in a deep-water setting.

shales or silty limestone strata. Outcrops of this formation have a characteristic stair-step profile in which the resistant limestone layers form ledges and low cliffs separated by intervening smooth slopes developed on the softer shale or silty limestone strata (Figure 7.10A, and Field Trip #2 for this chapter). Again, this cyclic pattern of limestone-shale-limestone-shale reflects the rhythmic oscillation of water depth in the basin. The limestone commonly contains

FIGURE 7.10. (A) Outcrops of the Pennsylvanian Ely Limestone at Grindstone Mountain, west of Elko, Nevada. Note the alternating ledges and slopes that illustrate the cyclic patterns of deposition. (B) Exposures of Mississippian and Pennsylvanian formations in Frenchman Mountain, southern Nevada.

chert as nodules and lenses, along with fossils of brachiopods, corals, and other invertebrates. There are also some thin, lenticular masses of conglomerate in parts of the Ely Limestone that probably represent submarine landslides from the rim of the basin or the remnants of the Antler highland to the west.

In the southern Great Basin, sediments of Pennsylvanian and Permian age accumulated in and near the Bird Spring Basin (Figure 7.2). This basin was trough-shaped and probably not as deep as the Ely and Oquirrh Basins to the north. It also seems to have received less land-derived sediment than those basins, suggesting that it was farther offshore or that major rivers from the interior of Pangaea did not discharge nearby. To the west, the Bird Spring Basin, or Trough, was bounded by the remnant Antler highlands, which by Permian time had been reduced to a series of islands. Farther offshore, lateral faults were slicing fragments of the ocean floor and shifting them laterally along the outer edge of western Pangaea. The Bird Spring was connected via a shallow platform to the deep Ely Basin on the north, prompting many Great Basin geologists to refer to these two sinks as the Bird Spring-Ely Basin.

Late Mississippian to Early Permian sediments deposited in the Bird Spring Basin vary in thickness from about 2,000 to more than 5,000 feet, suggesting variable rates of subsidence across the basin. This pattern may reflect a complex array of active faults that displaced the basin floor by varying amounts at different times and places. Sediment deposited in the Bird Spring Trough makes up the limestone of the Bird Spring Formation and Callville and Pakoon Limestones of southern Nevada (Figures 7.10B, 7.11). These carbonate rocks tend to include fine sand and shale in the southwestern Great Basin, which suggests proximity to land immediately east in what is now southwest Utah and adjacent Arizona (Figure 7.2). Some of the limestone layers in the Callville and Pakoon Formations contain granular deposits that appear to have been transported by wind during times of low sea level when carbonate dunes formed on the exposed shelf bordering the Bird Spring Basin. Farther offshore, massive sequences of cherty limestone with minor shale formed from sediments in the thick Bird Spring Formation (Figure 7.11). Several zones within these carbonate strata contain abundant fossils of normal shallow marine organisms such as brachiopods, bryozoans, and crinoids.

Beyond the western margin of the Bird Spring Basin, in what is now the Death Valley region, a complex pattern of deeper basins,

FIGURE 7.11. Outcrops of the Pennsylvanian Bird Spring Formation in the Arrow Canyon Range of southern Nevada. Inset photo illustrates the fragmentary fossils (mostly crinoids, brachiopods, and bryozoans) common in the limestone strata.

FIGURE 7.12. Volcanic rubble in the Permian Black Dyke Formation in the Pilot Mountains of western Nevada. Inset is a close-up view of gray sandstone comprised mostly of volcanic rock particles cut by white calcite veins.

linked to lateral displacement of crustal blocks, appears to have developed in Late Paleozoic time (Figure 7.2). In these basins, fine-grained mud and sand was transported by gravity flows, turbidity currents, and as suspended silt and clay from land. The Keeler Canyon Formation (Pennsylvanian to Early Permian) and the Owens Valley Group (Permian) of eastern California consist of mixed sediments of carbonate mudstone, siltstone and sandstone with graded bedding, conglomerate, and breccia. These sediments were deposited seaward of the shallow carbonate bank and the Bird Spring Basin by channelized debris flows and turbidity currents that moved down a relatively steep submarine slope from the shallow shelf to the east. At times of low sea level, rivers may have reached far enough west from the shoreline in Utah to scour channels in the exposed limestone seabed to deposit river gravels near the retreating coastal seas. Both the Owens Valley Group and the Keeler Canyon Formation exhibit cyclic patterns of sedimentation like those observed in the Pennsylvanian and Permian shallow-water deposits farther east, though less pronounced.

The Havallah Basin of northwestern Nevada was yet another site of deep-ocean sediment accumulation during the Late Paleozoic Era. However, because the accumulated materials were later displaced both east and west by Mesozoic thrusting, deformed and metamorphosed in the process, there is some uncertainty about the precise location, size, depth, and tectonic significance of this basin. Nonetheless, the sediments comprising what geologists describe as the Havallah sequence and related formations are clearly of deep marine origin. This package of rocks consists of a mixed assemblage of siltstone, sandstone, carbonate mudstone, and various volcanic rock types. Fossils are rare in this assemblage but do resemble North American faunas of the Late Paleozoic, suggesting that the basin was not far from the margin of Pangaea when it accumulated sediment. The volcanic materials include breccia, turbidites, and lavas (Figure 7.12) that might have been transported into the basin from a volcanic island arc to the west or could have been emplaced within the basin, perhaps both.

These volcanic components of the Havallah sequence have prompted some geologists to suggest that the basin may have opened by extension between an offshore volcanic arc and the eroded Antler highland to the east. Such basins are known as back-arc basins and result from the tensional forces associated with the outward migration, or "roll-back," of an offshore trench and the subducting plate beneath it (Figure 7.13). The precise arrangement of plates along this portion of the western margin of Pangaea is uncertain, so the interpretation of the Havallah basin as a back-arc basin is still debatable. Nonetheless, the complex assemblages of Pennsylvanian and Permian rocks in the western Great Basin clearly express steadily evolving plate tectonic interactions along the western margin of Pangaea during the final phases of the Paleozoic Era.

Late Paleozoic Life on the Great Basin Seafloor and the Permian-Triassic Extinctions

The Paleozoic fauna of marine invertebrates that became established in the Ordovician Period continued to thrive across the Great Basin seafloor during the plate tectonic convulsions of the Pennsylvanian and Permian. Brachiopods, bryozoans, corals, crinoids, ammonoid cephalopods, conodonts, and sponges remain plentiful as fossils in Pennsylvanian-Permian strata in the Great Basin (Figure 7.14). In addition, well-preserved fossils of echinoderms like modern brittle stars (class Ophiuroidea) have been discovered in the Pequop Formation of northern Nevada. However, the introduction of great quantities of land-derived sediment from the rising Ancestral Rocky Mountains (literally, a muddying of the waters), led to a decline in the abundance and diversity of organisms adapted to clear-water marine carbonate environments. Brachiopods, corals, and bryozoans became gradually less abundant in the Late Paleozoic Era, while organisms better suited to muddy water, such as gastropods and bivalve molluscs, became more common and diverse.

Of particular importance in Late Paleozoic strata are fossils of small unicellular organisms known as fusulinids. Though they were

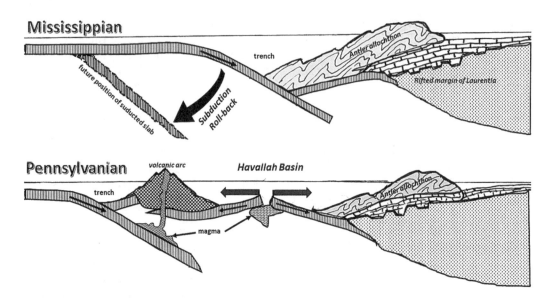

FIGURE 7.13. Possible origin of the Havallah Basin in Pennsylvanian time. The "roll-back" of a subducting slab results in tensional forces that open a back-arc basin landward of the volcanic island arc.

FIGURE 7.14. Common Pennsylvanian and Permian fossils from the Great Basin: (A) brachiopods from the Permian Gerster Formation, Confusion Range, Utah; (B) *Neozaphrentis*, a solitary coral from the Ely Limestone, Egan Range, Nevada; (C) crinoid columnals from the Ely Limestone, Egan Range. All scale bars except B = 1 cm.

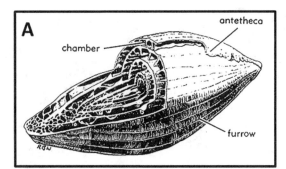

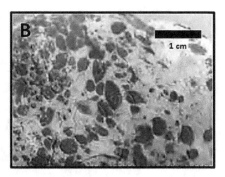

FIGURE 7.15. Test structure and fossils of fusulinids from the Great Basin: (A) form and interior structure of the fusulinid shells (called a test); (B) fuslinid fossils from the Ely Limestone of western Utah; (C) microscopic view of the internal structure of *Paraschwagerina elongata*, a fusulinid from the Bird Spring Formation of southeastern California. Image A from Kansas Geological Survey; image B from Utah Geological Survey Bulletin 133; image C from USGS Scientific Investigations Report 2013-5109.

single-celled, the cell was unusually large and had many specialized internal structures and organelles. More importantly, the fusulinids secreted a football-shaped shell, or test, composed of calcium carbonate and sometimes up to several millimeters in length. In rock outcrops, fusulinid fossils look like grains of rice or wheat of various colors. These shells, though small by human standards, are extraordinarily large for a unicellular creature and are preserved in great abundance in some intervals of the Ely Springs Limestone, Bird Spring Formation, and other Pennsylvanian-Permian rock units. Internally, the shells of fusulinids are very complex, with lengthwise coiling and numerous internal chambers arranged in an intricate pattern (Figure 7.15). The details of this internal structure are used by paleontologists to distinguish the many species of fusulinids from one another. Fusulinids first appear as fossils in Mississip-

pian strata. They diversified and grew larger in Pennsylvanian time and died out at the end of the Permian Period. These protozoans lived in a variety of habitats but appear to have been especially abundant in shallow, warm seas.

The Late Paleozoic seas of the Great Basin also harbored many types of fish. Isolated teeth of both cartilaginous and bony fish occur in several Permian and Pennsylvanian Formations of the Great Basin and nearby regions. However, the sandy limestone of the Permian Phosphoria Formation of northern Nevada and southern Idaho has produced the most interesting and varied assortment of fish remains. Small teeth, scales, and bone fragments of primitive bony fish known as palaeoniscoids occur in the Phosphoria strata of northeast Utah and adjacent Wyoming and Idaho. These fossils document the presence of several different types of bony fish in the Permian seas, but the fossil

FIGURE 7.16. Tooth whorl of *Helicoprion* from the Phosphoria Formation of southern Idaho. Image from the Idaho Virtual Museum, Idaho State University (license CC-BY-NC 4.0).

remains are too fragmentary to permit complete reconstructions. The remains of *Acrolepis*, a predatory, bony fish up to 10 feet long, have been identified from the Phosphoria Formation in southeast Idaho. The most intriguing fish fossils in the Phosphoria Formation are the spiral tooth whorls from a large, shark-like fish known as *Helicoprion* (Figure 7.16). Studying these odd structures from phosphate mines of southern Idaho, paleontologists have concluded that they represent the tooth whorls from a "buzz-saw shark" related to the modern ratfish. The spiral of more than 100 sharp teeth in the mouth is thought to have rotated backward when the jaws of *Helicoprion* closed around prey such as squid-like cephalopods or small fish. This motion impaled the prey and forced it farther back into the mouth where it was swallowed. Very little skeletal material is known from *Helicoprion*, suggesting that it was a cartilaginous fish. It likely possessed a streamlined, powerful

tail, and probably grew to a length of more than 20 feet. Such a swift and rapacious predator must have been near the top of the food chain in the Permian seas of western Pangaea.

Near the end of Permian time, about 252 million years ago, the greatest mass extinction in earth history, The Permo-Triassic extinction, began and ultimately led to an extraordinary decline in the abundance and biodiversity of life on a global scale (Figure 4.11). Paleontologists estimate that as much as 96 percent of all marine species in the Paleozoic fauna disappeared, while about three fourths of all land-living species vanished. This was the greatest biotic calamity in the nearly 4-billion-year history of life on our planet. Across the Great Basin seafloor, coral reefs died out, and the brachiopods, crinoids, bryozoans, ammonites, and gastropods that shared the reef habitat suffered devastating losses of biodiversity. The last of the trilobites and their arthropod relatives, eurypterids, or "sea scorpions," vanished forever. Not a single fusulinid, amazingly abundant in Pennsylvanian time, survived the Permo-Triassic extinction. The great die-off eliminated so many organisms from the Paleozoic fauna, an evolutionary assemblage that had characterized marine life for nearly 200 million years, that sea life would be permanently altered. In the geological periods following the Permo-Triassic extinctions, the few survivors of the ecological catastrophe would join newly evolved lineages to establish the Modern fauna of marine invertebrates (Figure 4.11). In this fauna, the bryozoans, brachiopods, and crinoids survive but with far fewer species and in more restricted habitats than their Paleozoic ancestors. We will explore the recovery from the Permo-Triassic extinction, and the transition to the Modern fauna, more fully in the next chapter, but it is worth considering what might have triggered the unprecedented collapse of marine diversity.

Throughout the late twentieth century, there were many hypotheses evoking a wide range of events sufficiently disruptive to stimulate such a massive annihilation of marine life. The pace and pattern of the extinction, along with evidence of contemporaneous geological and oceanic events, are critical in evaluating the proposed mechanisms. Our knowledge of these factors has improved dramatically over the past three decades, as new analytical techniques have been applied to the study of Permo-Triassic rock intervals and fossils. We now know much more about the pattern of the extinction, the conditions of seas, the global biogeography, and concurrent geological activity on land and in the ocean basins than just 30 years ago. Though certain aspects of the extinction are still puzzling, the new data have provided some powerful clues about the triggering mechanism, underscoring the unique severity of this event. Except for the current deterioration of marine ecosystems because of human activity, nothing like the Permo-Triassic extinction has ever affected life in the sea so severely.

Careful study of the fossils from rock sections spanning the Permo-Triassic boundary from around the world suggests that the extinction was a "sudden" event that may have swept through the world ocean in as little as 60,000 years. It also appears that both marine and terrestrial life were affected at about the same time. In the seas, sessile bottom-dwellers appear to have been more strongly affected than more mobile swimming creatures such as fish, nautiloids, and conodonts. This suggests that—perhaps—some event affected seafloor conditions first, allowing the swimming organisms to escape its effects, at least temporarily. Permo-Triassic boundary rocks include clay or black shale that typically exhibits a negative $\delta^{13}C$ excursion. This could signify an increase in the burial of organic carbon and/or reduced productivity of phytoplankton. Both are signs of a disrupted carbon cycle in the seas.

Leading up to the Permo-Triassic extinctions, several geologic and tectonic events might have created a unique set of biotic stresses in the marine ecosystem. The assembly of Pangaea through the amalgamation of continental masses greatly reduced the area of shallow continental shelves where the organisms in the Paleozoic fauna were concentrated. With less suitable habitat, competition for living space and resources would have been severe, and many groups of organisms might have died out in the battle for survival. However, as we have seen, the formation of Pangaea was a gradual process that played out over millions of years, and the ecological effects of this event would likely have been equally protracted. In addition, intense volcanic eruptions were in progress about that time. In what is now Russia, roughly a million cubic miles of lava erupted over the final million years of the Permian Period. These cooled lavas are known as the Siberian Traps, a vast region encompassing about 3 million square miles of northern Russia covered by basalt up to 10,000 feet thick. Such a voluminous eruption would have released enormous amounts of volcanic gases such as sulfur dioxide and carbon dioxide to the atmosphere. A single year could have released an estimated 1.5 billion tons of sulfur dioxide. The lava is thought to have ignited both living forests and coal deposits of Pennsylvanian age, creating more carbon dioxide, methane, and toxic ash. These gases would have acidified the seas, produced runaway greenhouse heating of the Earth, and potentially introduced toxic elements such as mercury, nickel, zinc, and lead into the seas.

Some climate models indicate that the Earth's surface temperatures rose by 14–18°F at the end of the Permian. The rapid heating, along with toxification of the surface waters, would have decimated plankton populations. In turn, an ensuing flood of organic matter to the seafloor would have consumed oxygen from the bottom waters, leading to suffocating anoxic conditions in the murky depths. The increased vertical temperature profile in the seas would likely have slowed oceanic circulation, preventing reoxygenation of bottom water and the dispersal of toxic substances. As the oceans stagnated, marine organisms died on massive scales, and land life began to wither under the rapidly warming and drying global climate. The death spiral on land and in the sea accelerated rapidly until the volcanic eruptions waned and clouds of volcanic ash hanging in the atmosphere began to slowly cool the planet.

Eventually, as the supercontinent of Pangaea began to break apart early in the Triassic Period, life on the Earth began to recover from the greatest biotic crisis it had ever faced. While the organisms in the Modern fauna were starting to restore biodiversity in the seas, tectonic events linked to the fragmentation of Pangaea began to lift the Great Basin region, expelling the seas and permanently converting the seafloor to land.

This final chapter in our story began even before the Permo-Triassic extinctions restructured life. As ecological drama played out at the end of the Paleozoic Era, the Great Basin seafloor was in the process of disappearing forever.

8

The Last Gasp of the Great Basin Seafloor

The Oceans of the West in the Age of Reptiles

The Permian Period, and concurrently the Paleozoic Era, ended about 252 million years ago. By this time, the supercontinent of Pangaea was fully assembled and surrounded by the Panthalassa Ocean (Figure 8.1). The Panthalassa Basin was probably underlain by several oceanic plates, but their precise arrangement is uncertain as much of the ancient seafloor has since disappeared into subduction zones. Still, there is good geological and geophysical evidence that numerous deep-sea trenches and volcanic arcs existed in this vast ocean, implying several zones of convergence and subduction in a mosaic of oceanic plates. Remnants of these arcs, the sediments of the intervening basins, and even small microcontinents were later thrust onto land in several portions of Pangaea, including the Great Basin region. Reconstructing the original position and pattern of these displaced masses of rock is a difficult task, but their presence is a clear indication that the Panthalassa Ocean Basin consisted of multiple plates with complex interactions, not unlike the plate tectonics of the western Pacific seafloor today.

The transition from the Paleozoic to Mesozoic Era was a time of fascinating and interconnected changes in global climate, geography, and life. As Pangaea formed during the Late Paleozoic Era, the global climate shifted from cool glacial conditions to a more varied seasonal and arid domain. By the beginning of Mesozoic time, vast deserts covered as much as 50 percent of interior Pangaea, while more temperate climates prevailed in the coastal regions. Summer temperatures in the deserts at latitude 30 degrees south might have routinely reached 115°F. Seasonal monsoons were strong in many areas of Pangaea, with regular flood events. These severe climatic conditions persisted until Middle Mesozoic time, when Pangaea began to break apart. The earliest phase of Pangaean fragmentation began in Triassic time, with the formation of a rift zone between modern northeast North America and northwest Africa. However, the main phase of separation began about 185 million years ago, in the Early Jurassic Period, when the nascent north Atlantic Ocean Basin began separating North America from Eurasia. It would take another 100 million years of additional rifting for the fragments of Pangaea to become recognizable as today's continents.

As Mesozoic time passed, the drift of continental masses was accompanied by the expansion of the modern ocean basins. A surge of volcanic activity at mid-oceanic ridges worldwide accelerated the rate of seafloor spreading. Global geography changed dramatically, exerting profound effects on climate and oceanic circulation. Carbon dioxide belched from undersea and terrestrial volcanoes and raised the levels of atmospheric carbon dioxide to a staggering 2,000 ppm (parts per million), occasionally spiking even higher. These volcanic greenhouse gases warmed the Earth, resulting in a persistent shift in all aspects of climate. By Cretaceous time, the world was ice free and, not surprisingly,

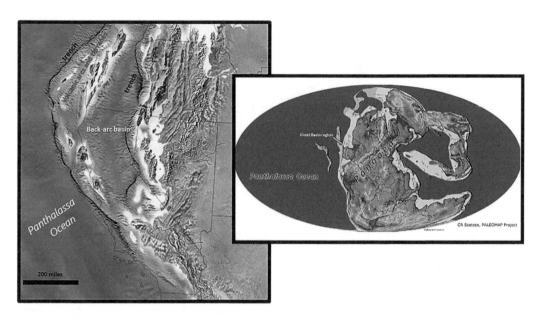

FIGURE 8.1. Western North America and global geography (*inset*) at the Permo-Triassic transition, 252 million years ago. North America base map prepared by R. Blakey, Colorado Plateau Geosystems Inc. Global map from C.R. Scotese (2014), Atlas of Permo-Triassic Paleogeographic Maps (Mollweide Projection), vols. 3 and 4, PALEOMAP Atlas for ArcGIS, PALEOMAP Project, Evanston, IL.

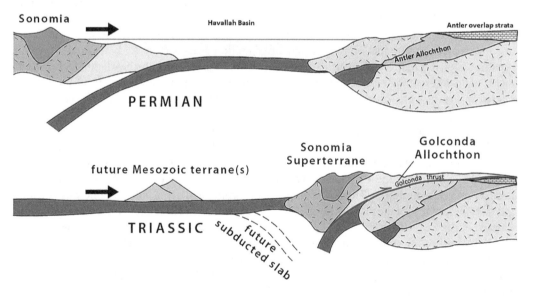

FIGURE 8.2. Illustration of a plausible plate tectonic mechanism for the Sonoma orogeny and accretion of the Sonomia superterrane in western Northern America in Late Permian time. Other explanations have been proposed and may be equally valid. Adapted from DeCourten and Biggar (2017), listed in Chapter 1 references.

sea level stood higher than at any time since the Proterozoic Era. In Late Cretaceous time, for example, sea level was some 1,000 feet above the level of modern oceans. Seawater was warm, and vertical mixing appears to have been stagnant, allowing anoxic bottom conditions to develop repeatedly in many regional basins.

All life was affected by these profound planetary changes. The ecological niches left vacant after the Permo-Triassic extinction, coupled with the era's rapid environmental shifts, stimulated an evolutionary burst that resulted in some of the most intriguing plants and animals in the long history of life on Earth. Dinosaurs on land, sea monsters in the oceans, the earliest mammals, giant fern-like trees, and the earliest flowering plants all appeared during the Mesozoic Era. As we will see, the remains of some these creatures, preserved in the Early Mesozoic sediments of the Great Basin seafloor, are quite different from those of the Paleozoic fauna.

Accretionary Tectonics of Western North America: The Beginning of the End for the Great Basin Seafloor

The formation of Pangaea, and the beginning of its subsequent breakup in the Early Mesozoic Era, appears to have intensified plate convergence along the western margin of the supercontinent. In what became western North America, volcanic and tectonic activity linked to convergent plate interactions were active in both Late Permian and Early Triassic times. Volcanic and exotic rocks of this age are known from northern California and northwest Nevada, suggesting the presence of several offshore island arcs and microcontinents at the dawn of the era. The volcanic rocks of the island arcs are of the type that forms from magma generated by subduction, a process that requires convergent interactions between oceanic plates.

Another indication of enhanced convergence was discovered in the 1960s by geologists studying the structure of the Sonoma Range in northwest Nevada. There, a sequence of deep-ocean rocks appeared to have been thrust eastward over the edge of the shallow continental

shelf during the Permian–Triassic transition. The basal thrust fault along which the deep-sea rocks were transported was named the Golconda thrust, and the displaced mass of rocks above it became known as the Golconda allochthon (Figure 8.2). The thrusting of the Golconda allochthon was thought to have been linked to the Sonoma orogeny, a period of mountain-building named for the mountains where the evidence of this disturbance was first detected in the rock record. Deformation, displacement, and orogeny were like the Antler orogeny but occurred father west. The rocks within the Golconda allochthon consisted of bedded chert, impure limestone, conglomerate, argillite, and volcanic rocks that had originated in a deep-ocean basin west of the continental margin. This basin may have been the Havallah basin or perhaps some other more distant deep-sea trough that converged toward North America from the west in Permian and Triassic time.

Several explanations for the triggering mechanism have been proposed for the Sonoma orogeny, but most involve some sort of plate convergence along the western edge of North America. A popular model for this orogeny involves the approach of a volcanic arc or microcontinent from the west that squeezed deep-sea rocks from the Havallah basin eastward over the continental shelf. But evidence also exists for lateral movement of plates (from the north or south) along the continental margin during the Early Mesozoic time frame. A combination of convergent and transform interactions, where plates come together obliquely and slide past one another while they collide, is also possible. Such interactions generate forces known as "transpression," which produce a complex set of tectonic consequences. What is evident from the Permian-Triassic rock record of the western Great Basin is that multiple plates converged against the western margin of North America as it began to separate from Pangaea in Early Mesozoic time.

The Golconda allochthon was transported some 40 miles east over the western portion of the older and mostly subsided Antler allochthon. This motion lifted the mass of deformed

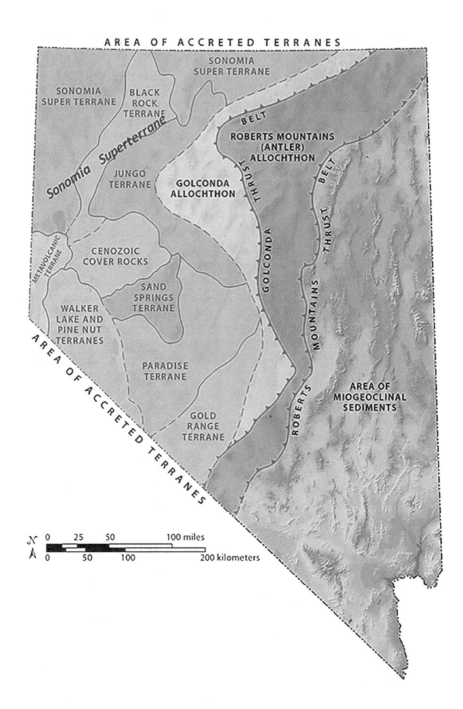

FIGURE 8.3. Mesozoic accreted terranes of the western Great Basin. Terrane locations and boundaries are approximate, and the Sonomia superterrane may encompass several of those depicted. Adapted from DeCourten and Biggar (2017), listed in Chapter 1 references.

rock and elevated it as land beginning in the Late Permian Period. Not long after the Golconda allochthon emerged, additional masses of rock transported on oceanic plates approached the western margin of North America. Some of these plates carried volcanic island arcs, while others bore microcontinents previously rifted from the margins of Pangaea or reef-like masses of carbonate rocks. In Triassic and Jurassic time, each of these relatively small masses was crushed against the continental edge, adding to the basement material already in place. Geologists refer to these masses of exotic rocks as terranes.

The pattern of accreted terranes in the western Great Basin is complex and still somewhat uncertain because the rocks were commonly deformed and metamorphosed as they were attached to Mesozoic North America. Some of the terranes were split and dismembered as they docked, with fragments embedded into the continental margin at different places and from different directions. To further complicate matters, the approaching terranes could collide with each other, becoming amalgamated into composite terranes before they reached the margin of North America. One such composite terrane, or "superterrane," as some geologists prefer, has been named Sonomia, after the Sonoma orogeny that may have been initiated by its accretion. Composite terranes are particularly problematic because they consist of rocks of diverse ages and origins and may have been deformed multiple times, even before they arrived in the Great Basin region. Given that the pattern and sequence of accreted terranes is such a bewildering mosaic, it is not surprising that the details of the Sonomia superterrane are still a bit controversial (Figure 8.3).

While this terrane accretion began in Permian time, it accelerated into the Early Mesozoic Era. This time frame coincides with the breakup of Pangaea and the isolation of the North America plate via the opening of the north Atlantic Ocean Basin. Seafloor spreading at the mid-Atlantic ridge propelled North America westward to actively meet the oncoming, terrane-bearing plates. The oceanic lithosphere

approaching the western margin of North America was consumed in an east-dipping subduction zone. Each time a mass of rock was added to the western margin of North America, the Great Basin seafloor was deformed and lifted, and the seas that formerly covered it withdrew from the rising land. By Late Jurassic time, about 150 million years ago, most of the accreted terranes were in place, and the seafloor was exposed as new land with a kaleidoscopic pattern of exotic rocks. The continental margin now extended westward nearly to California, close to its modern position. After more than 500 million years under ancient oceans, the Great Basin seafloor was gone forever. (Or is it? See the epilogue at the end of this chapter.)

Triassic Rocks of the Great Basin

After the Sonoma orogeny and the accretion of the Sonomia terrane, a period of erosion followed, marked in the rock record as a widespread regional unconformity separating earlier Permian from Triassic strata. In the southern Great Basin, rivers flowing from the Colorado Plateau region deposited gravel on the eroded surface of Permian limestone (Figure 8.5B). Then, the deformed crust subsided enough to allow marine sediments to once again accumulate in shallow seas and bays adjacent to land in west-central Nevada (Figures 8.4, 8.5A). By Middle and Late Triassic time, silty limestone was deposited in interconnected, shallow basins stretching from south-central Nevada to southeast Idaho. In the western Great Basin, the sediments appear to have been deposited in deeper water (Figure 8.5C, D) and commonly contain volcanic materials derived from islands or terranes from the west. In southernmost Nevada and adjacent southwestern Utah, the open marine sediments grade landward into mudflat and river-deposited sediments in the Colorado Plateau region.

Across most of the Great Basin, the strata immediately above the Permian-Triassic unconformity contain conglomerate generated during exposure and erosion of the underlying Permian rocks. The Early Triassic shoreline was situated in central Utah, where rivers deposited mud

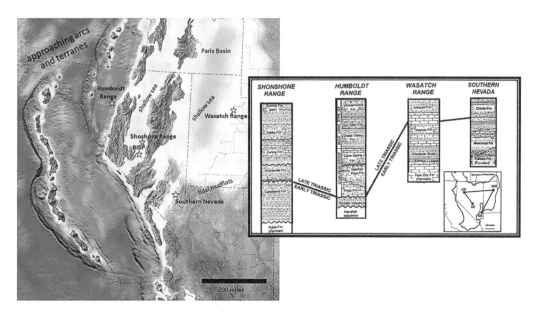

FIGURE 8.4. Early Triassic paleogeography of the Great Basin region and general Triassic stratigraphy (*inset*). In addition to the rock units identified in these columns, many other formations have been named across the region. Paleogeographic base map prepared by R. Blakey, Colorado Plateau Geosystems Inc.

FIGURE 8.5. Triassic sedimentary rocks of western Nevada: (A) shallow-water carbonate sediments in the Virgin Limestone near Blue Diamond; (B) river-deposited conglomerate in the Timpoweap Member, Moenkopi Formation, southern Nevada; (C) rubble from undersea landslides in the Humboldt Range, northwest Nevada; (D) interbedded deep basin sandstone and siltstone in the Sand Springs Range, western Nevada.

and sand comprising the rocks of the Moenkopi Formation (or Group, as some geologists prefer) in southern Nevada and the Ankareh and Woodside Formations of northern Utah. These coastal sediments interfinger with marine limestone offshore (to the west), indicating that the shoreline migrated in and out repeatedly across the eastern portion of the Great Basin seafloor. Footprints and fragmentary fossils from the terrestrial strata of the Moenkopi Formation suggest that several types of reptiles and amphibians inhabited the coastal plains. Offshore, to the west, impure carbonate sediments accumulated in an open marine setting. The limestone of the Thaynes Formation and Virgin Limestone Member of the Moenkopi Formation contain fossils such as cephalopod and bivalve molluscs that indicate a shallow marine environment. In western Nevada, volcanic materials in Early Triassic strata, such as the Koipato Formation (or Group) and the Candelaria Formation, signify eruptions that occurred in oceanic or near-oceanic settings. The presence of active volcanoes along the western margin of North America was likely related to subduction of the oceanic plates that accompanied terrane accretion there.

By Middle and Late Triassic time, thick successions of limestone and siltstone, sometimes accompanied by volcanic components, accumulated in the isolated, but presumably interconnected, basins across the Great Basin. The Luning and Gabbs Formations of western Nevada are collectively more than 10,000 feet thick, while in the Humboldt Range area the Star Peak Group (Prida/Favert, Augusta Mountain/Natchez Pass, and Cane Spring Formations) and Auld Lang Syne Groups (Grass Valley, Dun Glen, and Winnemucca Formations) exceed 5,000 feet. The formations of the Star Peak Group are dominated by fossiliferous carbonate sediments deposited in a shallow-water setting near the edge of the continental shelf. Just west of the shelf, the seafloor descended into a deeper basin that occasionally received large masses of shelf rubble likely dislodged by earthquakes or storms. The carbonate sediments deposited at the edge of the shelf commonly exhibit evidence

of downslope movement shortly after deposition (Figure 8.6).

The rocks in the various formations of the Auld Lang Syne Group are primarily muddy siltstone and sandstones mixed with less abundant limestone strata. Most of the upper part of the Auld Lang Syne Group was deposited on, or just offshore of, a muddy delta receiving fine-grained sediments from Late Triassic rivers draining the region of the modern Colorado Plateau to the southeast. By the end of the Triassic Period, the deltaic complex, following the retreating seas, had grown outward as far as western Nevada (Figure 8.7). At that time, around 200 million years ago, a broad, coastal plain then covered almost all the former Great Basin seafloor. Though some marine inlets and bays persisted into the Early Jurassic Period, the Late Triassic burial of the ancient seafloor under terrestrial sediments, coupled with continuing Mesozoic uplift, expelled the seas from the Great Basin region for the final time.

Life in the Triassic Seas: A Great Recovery and Swarms of Sea Monsters.

Immediately following the ecological devastation of the Permo-Triassic, the earliest Mesozoic seas in the Great Basin were populated by a scant community of creatures hardy enough to have survived the crisis. Fossils and sedimentary structures from the earliest Triassic Virgin Limestone Member (of the Moenkopi Formation), most of the Union Wash Formation in eastern California, and the lower part of the Thaynes Formation (or Group) of western Utah are dominated by trace fossils, sponges, and microbial structures such as wavy laminations, thrombolites, and stromatolitic lumps. Crystalline, inorganic calcium carbonate cement reported in some portions of these formations indicates that the bottom waters were alkaline and anoxic. The rock record suggests that some areas on the seafloor were still somewhat hostile to life for about the first 5 million years of the Triassic Period.

However, other portions of the Great Basin seafloor appear to have been the scene of a remarkably rapid ecological recovery and were

FIGURE 8.6. Outcrop view of deformed sediments in the Triassic Prida Formation, Humboldt Range, Nevada. The carbonate mud and silt in this exposure slid down the sloping seafloor near the edge of the continental shelf prior to complete lithification.

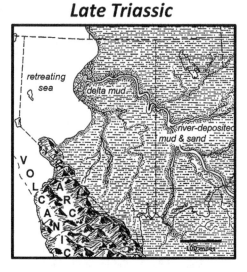

Early Triassic

shallow
sea

Sonoma Highlands

coastal
plain

(mudflats)

edge of North America

developing
volcanic arc

Colorado River

100 miles

Late Triassic

retreating
sea

delta mud

river-deposited
mud & sand

VOLCANIC ARC

100 miles

FIGURE 8.7. Withdrawal of the Triassic seas in the Great Basin. Deltaic and river-deposited sediments accompany the westward shift in the shoreline by Late Triassic time. Modified from DeCourten (2003), listed in Chapter 1 references.

quickly populated by a profuse array of organisms within just a few million years of the great extinction. One such place was the area just west of Paris in southeast Idaho, very near the northern boundary of the Great Basin. Here, from rocks of the lower Thaynes Formation (or Group) of Early Triassic age, a remarkably rich association of fossils has been recently discovered (see Brayard et al. [2017], Chapter 8 references). These fossils represent a robust community of invertebrates, vertebrates, and algae known collectively as the Paris biota. The Paris biota included many types of sponges, bivalves, gastropods, shrimp and lobster-like crustaceans, brachiopods, conodonts, ophiuroids (brittle stars), crinoids, ammonoid cephalopods, nautiloid cephalopods, fish, and algae. This is an astounding diversity, considering that the great extinction occurred just a few million years earlier. Paleontologists are still debating the pace and pattern of the Triassic recovery from the end-Permian extinction, but it appears that life on the seafloor staggered back in several pulses before breaking into a full sprint of biodiversity after Middle Triassic time. The Paris biota demonstrates an explosive recovery where conditions were favorable.

By about 245 million years ago, multiple lineages of marine invertebrates were evolving rapidly to fill the ecological niches vacated by their extinct ancestors. This great burst of evolution includes groups familiar to us today: gastropods (snails), bivalves (clams and oysters), squid, sea urchins, and modern corals. Because the Triassic marine assemblage includes so many familiar forms, it is considered to represent the rise to dominance of the Modern fauna of marine invertebrates (Figure 4.11). Recent studies have suggested that the shift to mollusc-dominated communities may have begun in the Late Paleozoic Era, but it is in the Triassic Period that the change becomes notable in the Great Basin strata. The bivalve molluscs evolved many new lineages, replacing the Paleozoic brachiopods as the most abundant seafloor shellfish. Triassic clams developed specializations for attachment to the seafloor and burrowing into the soft sediment. These innovations led to

rapid diversification of the various bivalve lineages. Gastropod molluscs also flourished in the Middle Triassic seas, but their fossils are less common than the bivalves in most exposures of Triassic strata.

After the extinction of the tabulate and solitary corals of the Late Paleozoic, a new group of corals, the scleractinians or "stony corals" emerged in the Triassic to assume, along with the calcareous sponges, the role of primary reef-builders in the Mesozoic seas. Though the scleractinian corals belong to the same higher taxonomic categories as the Paleozoic corals (phylum Cnidaria, class Anthozoa), their anatomy and skeletal structure are sufficiently distinctive to suggest only a distant evolutionary relationship with earlier types. The scleractinian corals probably evolved from a noncoral Cnidarian ancestor but came to mimic the very successful reef-building habits of their Paleozoic predecessors. At least 18 different species of corals have been discovered in the Luning Formation of western Nevada (Figure 8.8), where reef-like masses up to 500 feet long and 50 feet high grew in shallow water at the edge of the Triassic continental shelf. Many of the Triassic coral colonies have a flattened, plate-like shape that seems well-designed to absorb sunlight in the illuminated part of the water column. This may suggest that, like their modern descendants, Triassic corals had a symbiotic relationship with photosynthetic algae. Fossils associated with the coral-sponge reefs include bivalve molluscs, brachiopods, crinoids and other echinoderms, and even marine reptiles. Once established, the scleractinian corals maintained their reef-building prominence to the modern time.

As the evolutionary outburst swept through the Triassic seas, one of the invertebrate groups most affected were the ammonoid cephalopods. Fossils of these swimming predators are found in many of the Great Basin Triassic limestone formations, sometimes in spectacular abundance. These relatives of the modern squid and cuttlefish evolved rapidly, became widespread, and secreted robust, coiled shells that were preserved in many depositional environments.

Daonella rieberi, a bivalve mollusc from
the Prida Formation, Humboldt Range, Nevada

Polytholosia sp., a sponge from the
Luning Formation, Pilot Range, Nevada

Montlivalta marmorea, a solitary coral from the
Luning Formation, Pilot Range, Nevada

Pamiroseris norica, a colonial coral from the
Luning Formation, Pilot Range, Nevada

FIGURE 8.8. Non-ammonoid invertebrate fossils from Triassic strata of central and western Nevada. Scale bar = 1 cm in all figures. Specimens from USGS Professional Paper 1207 and the Keck Museum, University of Nevada, Reno.

Such attributes make ammonoid fossils extremely useful in subdividing and correlating Triassic rock sequences across the Great Basin and globally. Hundreds of species have been identified from Great Basin strata (Figure 8.9), each with distinctive features such as degree of coiling, ornamentation, and spacing and shape of the internal walls (septa). Their diversity notwithstanding, most ammonoids appear to have been active predators that acquired their prey with tentacled arms and ingested it through a mouth equipped with a beak (radula) and tooth-like spikes and plates. Ammonoids probably preyed on sessile bottom-living organisms such as gastropods, bivalves, arthropods, worms, and echinoderms, as well as swimming creatures such as small fish, conodonts, and other cephalopods.

The Great Basin Triassic oceans also teemed with fish. Fish fossils of this age are widespread and have been collected from numerous rock units including the Candelaria Formation of western Nevada, the Prida and Favert Formations of north-central Nevada, and the Thaynes Formation of northeast Nevada, Idaho, and Utah (Figure 8.10). The array of Triassic fishes is incredibly diverse and includes many types of bony fish (osteichthyans), cartilaginous fish (chondrichthyans), and lobe-finned fish (sarcopterygians). Within this assemblage are fish that appear to have played several different ecological roles and occupied multiple trophic

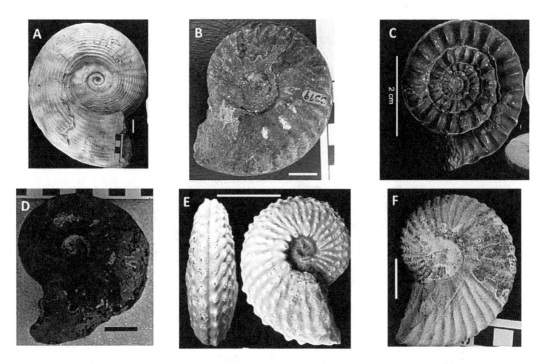

FIGURE 8.9. Ammonoid fossils from the Triassic Thaynes, Prida, and Luning Formations of the Great Basin region: (A) *Euflemingites cirratus*; (B) *Gymnoceras russeli*; (C) *Columbites crassicostatus*; (D) *Meekoceras tuberculatum*; (E) *Eoprotrachyceras subasperum*, lateral and ventral views; (F) *Achrochordiceras* sp. All scale bars = 2 cm. A, C, and E from Jenks et al. (2007); B and F from the Keck Museum, University of Nevada, Reno; D from the Natural History Museum of Utah.

levels in the marine food web. There are large predators, smaller prey fish, and even some that could live in both seawater and freshwater settings. Such variety appears in the earliest Triassic fish faunas, suggesting that fish either suffered less than most invertebrates during the Permo-Triassic extinctions or their recovery was amazingly rapid. Another interesting aspect of the Triassic fish fauna of the Great Basin is that it includes several species adapted to shallow tropical seas that are very similar, if not identical, to fish of the same age found in China and Japan. In the Triassic, Asia was located on the opposite side of Pangaea from the Great Basin, the two regions separated by the vast and deep Panthalassa Ocean. The co-occurrence of similar fish indicates the ability of at least some Triassic fish to migrate long distances. Perhaps these migrant fish used the shallow seas surrounding the island terranes scattered through the eastern Panthalassa Ocean as steppingstones to the west coast of Pangaea.

Saurichthys (Figure 8.10) was one of the most common predators in the Triassic seas of the Great Basin. Fossil remains of this slender fish have been found in the Thaynes Formation of northeast Nevada and southeast Idaho as well as in the Favert Formation of western Nevada. *Saurichthys* grew up to 4 feet in length and had a torpedo shaped body propelled by a powerful tail fin. Its pelvic, dorsal, and anal fins were located near the tail, and its elongated skull tapered into a narrow, pointed snout whose jaws were lined with small, spike-like teeth that seem well-designed to snap up swimming prey. Such specializations suggest that *Saurichthys* was an "ambush" predator that lurked in the crevices and nooks of the Triassic coral reefs, rocketing from the shadows to capture even the most elusive Triassic swimmers.

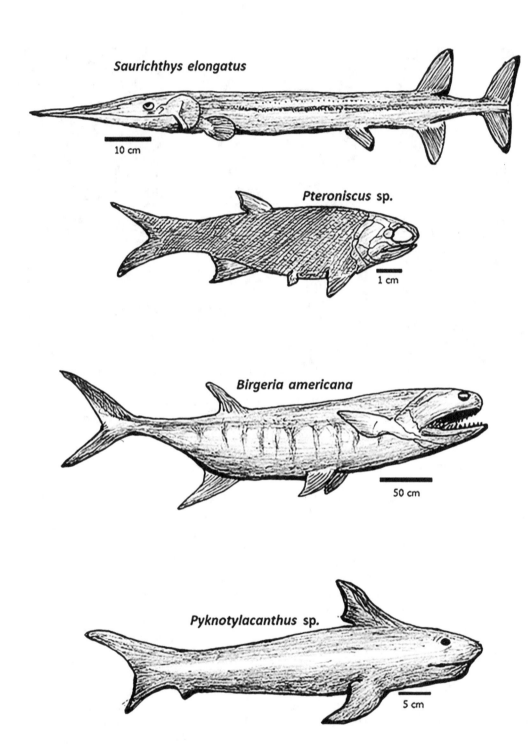

Saurichthys elongatus

10 cm

Pteroniscus sp.

1 cm

Birgeria americana

50 cm

Pyknotylacanthus sp.

5 cm

FIGURE 8.10. Examples of Triassic fish of the Great Basin region. These reconstructions are mostly based on fragmentary fossils and represent the likely, but somewhat uncertain, appearance of the fish in the Triassic seas of western North America. Note size variation.

Saurichthys was probably one of the top predatory fish of its time, but there were many other voracious marauders cruising the Great Basin seafloor then. In 2017, a new species of Triassic fish, *Birgeria americana*, was identified from fossils found in the Thaynes Formation of Elko County, Nevada (see Romano et al. [2017], Chapter 8 references). *Birgeria* was a large osteichthyan with an adult body length of about six feet (Figure 8.10). It had several parallel rows of sharp teeth about half an inch long lining robust jaws. Paleontologist have suggested that *Birgeria* consumed its prey like modern sharks: a single, powerful bite followed by a gulp of the unfortunate prey. *Saurichthys* and *Birgeria* were probably nonselective predators that consumed a mixed diet of ammonoids and other swimming cephalopods, smaller marine reptiles, and other fish. Many other species of chondrichthyan fish, such as *Pyknotylacanthus* and smaller osteichthyans like *Pteronisculus* (Figure 8.10), have been documented among the fossils collected from Great Basin Triassic rocks, and new ones are continuously discovered. The fossil evidence leaves no doubt that the fish fauna of those seas was diverse and included many different well-adapted lineages. However, even the largest predator fish had something to fear. Some 240 million years ago, some of the largest and most terrifying aquatic predators in the history of life invaded the Great Basin seas.

Sea Serpents of the Sage:
Great Basin Triassic Marine Reptiles

The Mesozoic Era is known worldwide as the Age of Reptiles for good reason. Most of the major reptile lineages first appeared on land during the Triassic Period, rapidly displacing the therapsids ("stem-mammals") as the dominant terrestrial tetrapods. Continuing evolution among Mesozoic reptiles would ultimately result in a dazzlingly diverse array of vertebrates that includes dinosaurs, birds, pterosaurs, turtles, crocodiles, snakes, lizards, and many other less-familiar and now-extinct lineages. The explosive terrestrial evolution of reptiles following the Permo-Triassic extinctions is one of the most pivotal events in the history of life on our

planet. This great burst was the result of numerous factors, but the ecological niches left vacant by the preceding extinctions were an important influence. The multiple open niches provided evolutionary opportunities previously unavailable to early reptiles. And the reptiles responded in dramatic fashion, diversifying into many specialized lineages to become the dominant vertebrates of Mesozoic Earth. But while dinosaurs are famous for the reptile takeover on land, there was an equally profound, almost simultaneous, transition to reptile dominance in the seas. It would produce some of the most fascinating creatures to ever swim the oceans covering the Great Basin seafloor.

Adaptation to the marine environment was one of the most remarkable aspects of the reptile radiation in the Early Mesozoic Era. The transition from life on land back into the oceans required significant modifications in the axial skeleton, limbs, feet, skull, ribs, lungs, diet, sensory systems, and reproductive strategies of terrestrial tetrapods. These evolutionary trends appear to have begun in some reptiles in the Late Permian. However, the fossil evidence suggests that less than 10 million years after the end of Permian time, several different lineages of well-adapted marine reptiles were thriving in the Triassic seas—an extraordinarily rapid transition. So varied are the Triassic marine reptiles that it seems unlikely they evolved simultaneously from a single terrestrial ancestor. Instead, it appears probable that several different groups of land-living reptiles adapted to the marine environment in rapid succession during the Early Triassic Period. By Middle and Late Triassic time, there were at least three main groups of marine reptiles in the Great Basin seas: the sauropterygians, the ichthyosaurs, and the thalattosaurs.

Sauropterygians (class Reptilia, order or superorder Sauropterygia) are among the earliest reptiles to appear in the Triassic seas and a persistent and diverse group that survived to the end of the Mesozoic Era in the form of the long-necked plesiosaurs. The pectoral girdle in the shoulder region of these reptiles was modified to allow powerful forward and backward

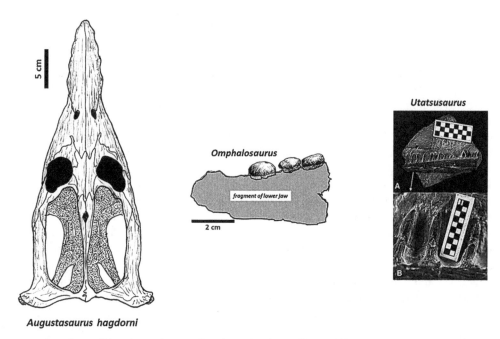

FIGURE 8.11. Some Triassic marine reptiles from the Great Basin: (A) skull of *Augustasaurus hagdorni*, Augusta Mountains, central Nevada; (B) teeth of the primitive Triassic ichthyosaur *Omphalosaurus*, based on a specimen from Germany; (C) *Utatsusaurus* from the Prida Formation, Humboldt Range, Nevada. Skull (A) based on reconstruction of Rippel et al. (2002); photo of Utatsusaurus jaw and teeth from Kelly et al. (2016), listed in Chapter 8 references.

strokes of the elongated forelimbs, the main source of propulsion. In later, more advanced sauropterygians, the hindlimbs were similarly modified as flippers to assist in swimming. The skulls of the sauropterygians are relatively small, with many relatively blunt, fang-like teeth projecting from the lower and upper jaws. The sauropterygians generally had elongated necks to move their small heads in search of prey on the seafloor or swimming nearby.

The best-known sauropterygian from the Great Basin is *Augustasaurus hagdorni*, the remains of which were described in 1997 from the Middle Triassic Favert Formation of the Augusta Mountains in west-central Nevada. This reptile was about 8 feet long, with an elongated neck and appendages modified into paddle-like flippers. The skull was about 15 inches long with a narrowed snout that would have moved through the water with ease to snap up fish or molluscan prey (Figure 8.11). The bones of *Augustasuarus* were found in dark-colored, lami-

nated mudstone likely deposited in relatively deep and calm water with minimal oxygen levels. However, a mass of coarse sandstone was found between the ribs of the partial skeleton. Such coarse granular sediment would have been unlikely to accumulate in the deeper offshore basin where the bones were preserved. This sand was probably ingested from the shallow seafloor farther east where near-shore sand deposits were more common. The sand might have been swallowed inadvertently while the reptile was feeding on bottom-living prey. However, the later and more advanced sauropterygians known as plesiosaurs commonly ingested stones for buoyancy control, and it is possible that *Augustasaurus* may have had a similar habit but without access to large stones.

Ichthyosaurs were a fascinating and diverse group of marine reptiles in the Triassic seas worldwide. These large, predatory reptiles were as long as 65 feet or more, weighed as much as 40 or 50 tons, and occupied the top tier in the marine food chain. The ichthyosaurs were the ecological equivalent of modern orcas, or killer whales, but included some species that were substantially larger. Their long, streamlined bodies resembled those of modern dolphins, though their powerful tail fins were vertical. They used their paddle-like appendages to maneuver as they pursued prey, which was almost anything that swam through the Late Triassic seas, including even the largest fish, other marine reptiles, turtles, and ammonites. Ichthyosaurs had large eyes and excellent vision, allowing them to hunt in murky or deeper water where their prey might attempt to evade the fearsome teeth and jaws. These reptiles also gave birth to live offspring, another factor behind their remarkable success in the Triassic seas. Each younger ichthyosaur was a miniature, fully functional, replica of the adult. Remains of ichthyosaurs have been found in Triassic strata of every continent except Antarctica, demonstrating their unparalleled success. These remarkable creatures, which rapidly developed from terrestrial ancestors in the Early Mesozoic Era, represent a magnificent story of adaptation and evolution in the prehistoric seas until their extinction in Middle Cretaceous time. More than 200 different species have been described by paleontologists, though the validity of some of these is doubtful because they were based on scrappy fossils or may have been sexual or age variants of a single species.

Several ichthyosaurs have been discovered in the Middle and Late Triassic rocks of the central and western Great Basin, primarily in the limestone and calcareous siltstone of the Prida, Favret, and Luning Formations. These formations consist primarily of calcareous sediment deposited in lagoons, bays, and inlets. Ichthyosaur remains commonly occur in association with the shells of bivalves, gastropods, corals, and ammonoids. However, the excellent swimming and diving abilities of ichthyosaurs would certainly have allowed them to move into deeper water in the western extremity of the modern Great Basin region. The concentration of ichthyosaur fossils in shallow marine deposits probably reflects conditions favorable for preservation as well as habitat preferences. A mass-mortality accumulation at Berlin-Ichthyosaur State Park in Nye County, Nevada, is one of the most spectacular concentrations of vertebrate fossils in the world and suggests some unusual circumstances. This site is discussed in more detail in the supplemental field guide available online.

One of the first ichthyosaurs discovered in the Great Basin is *Omphalosaurus nevadanus* from the lower portion of the Prida Formation in the Humboldt Range of northwestern Nevada. In 1902, J. C. Merriam of the University of California found a partial skeleton, which he scientifically described in 1906. The incomplete fossil material was not immediately identified as the remains of an ichthyosaur but was eventually assigned to that group of marine reptiles. Compared to other ichthyosaurs, the most distinctive aspect of *Omphalosaurus* is its small (approximately half inch diameter), button-like teeth (Figure 8.11). These teeth are quite unlike the sharp, slashing teeth of other ichthyosaurs and seem much better designed for crushing thin-shelled invertebrates such as ammonoids, some bivalves, crustaceans, and brachiopods. In addition to their rather odd shape, the teeth of *Omphalosaurus* were arranged in the jaws as an irregular mosaic, rather than set in linear grooves or sockets as with most other ichthyosaurs. A second species of *Omphalosaurus*, *O. nettarhynchus*, was described from the Prida Formation in 1987, but neither of these ichthyosaur species appear to have been particularly large. Because the fossil material is incomplete, the exact lengths of the Nevada specimens of *Omphalosaurus* are not known, but they were probably less than 4 feet long.

Another relatively primitive ichthyosaur, *Utatsusaurus*, was described in 2016 from the lower member of the Prida Formation and in the same general area where the remains of *Omphalosaurus* were discovered a century earlier. This ichthyosaur may be the oldest known from the Great Basin and was identified based on a jaw fragment with numerous teeth. More complete specimens of *Utatsusaurus* from Japan suggest it was somewhat lizard-like in

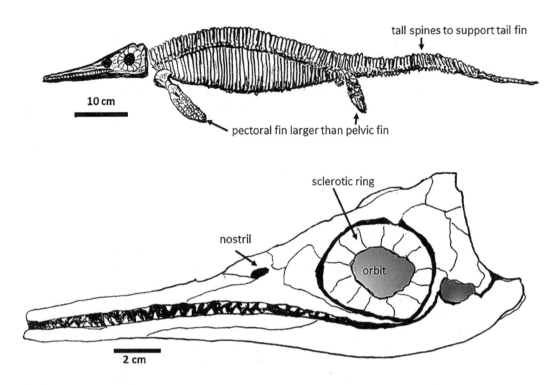

tall spines to support tail fin

10 cm

pectoral fin larger than pelvic fin

sclerotic ring

nostril

orbit

2 cm

FIGURE 8.12. Skeleton and skull of *Mixosaurus*, a small Triassic ichthyosaur from the Great Basin.

appearance and probably less than 3 feet long. The conical teeth of *Utatsusaurus* are about a half inch tall, heavily striated, and set in sockets aligned along a groove in the lower jaw (Figure 8.11). This small ichthyosaur most likely fed on fish, ammonoids, and squid-like cephalopods. The form of the teeth and other dental characteristics of the Nevada fossil are almost identical to the Japanese specimens and are the basis for the identification of this genus in Nevada.

Remains of another small ichthyosaur, *Mixosaurus*, have been collected from the Middle Triassic Favert Formation of the Augusta Mountains and the equivalent Prida Formation of the Humboldt Range in west-central Nevada (Figure 8.12). Paleontologists recognize at least two species, *M. nordenskioeldii* and *M. callawayi*, in the fossil material from these localities but both share many common traits. In general, *Mixosaurus* was probably 2–5 feet long and had large, flipper-like pectoral fins and smaller pelvic fins nearer the tail (Figure 8.12). The long, flexible tail probably bore a ridge-like fin and

helped *Mixosaurus* maneuver through the water by undulating the tail from side to side. The elongated, narrow snout reduced hydrodynamic resistance, and dozens of small teeth lined the upper and lower jaws. Embedded in the relatively large eye of *Mixosaurus* was a bony doughnut, the sclerotic ring, constructed of numerous articulating plates of bone. The sclerotic ring helped the eye retain its shape when rapid swimming or deep diving might otherwise have deformed the soft tissues. The large size and well-developed sclerotic rings common to most ichthyosaurs suggest these reptiles had excellent vision, which they utilized to locate prey under all conditions of water clarity, illumination, and depth. The nostrils were located a short distance in front of the eyes and allowed *Mixosaurus* to catch its breath when surfacing, like a small, reptilian version of a dolphin. Another small Triassic ichthyosaur, *Phalarodon fraasi*, was identified in 1910 from fragments of a fossil reptile skull and jaws also collected in the Humboldt Range. The fossils are so close to

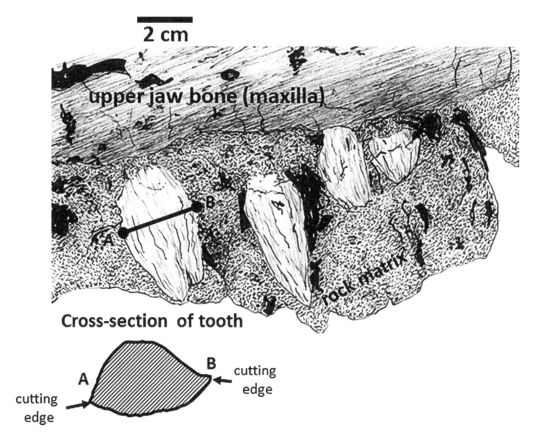

FIGURE 8.13. Teeth and upper jaw of *Thalattoarchon saurophagis* from the Favert Formation, Augusta Mountains, Nevada. Based on specimen photograph of Fröbisch, et al. (2013), listed in Chapter 8 references.

those of *Mixosaurus* that most paleontologists now consider them the same ichthyosaur genus.

The ichthyosaurs discussed thus far were all relatively small marine reptiles, rarely exceeding 4–5 feet long. But some ichthyosaurs known from the Great Basin were gigantic and would have been a terrifying sight. *Thalattoarchon saurophagis*, for example, was at least 28 feet in length and possessed large, blade-like teeth with two sharp cutting edges (Figure 8.13). These teeth, up to 5 inches long, were likely used to capture and slice prey that included smaller ichthyosaurs. The species name, *saurophagis*, or "reptile eating," suggests the presumed diet of this sea monster. The remains of *Thalattoarchon* were discovered in the Favert Formation of the Augusta Mountains in central Nevada.

However, only a partial skeleton and skull were preserved, leaving some uncertainty about its precise appearance. It does seem to be similar in appearance to *Cymbospondylus*, a larger and much better-known Great Basin ichthyosaur.

Cymbospondylus, a truly gigantic beast, was the first ichthyosaur described from Great Basin fossil material. In 1868, the illustrious paleontologist Joseph Leidy established this ichthyosaur genus based on fossil material from Triassic strata in the Humboldt and New Pass Ranges in northwestern Nevada. Leidy identified two species. In the early 1900s, scientists from the University of California collected new fossil material, identifying several additional species. Subsequent discoveries of *Cymbospondylus* fossils in such widely separated localities as

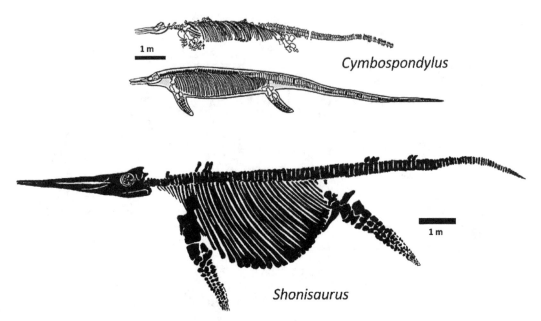

1 m

Cymbospondylus

1 m

Shonisaurus

FIGURE 8.14. Large ichthyosaurs from the Great Basin: *Cymbospondylus* skeleton and reconstruction (*upper*) and restored skeleton of *Shonisaurus* (*lower*).

Switzerland and China led to the recognition that the genus *Cymbospondylus* had a global distribution in Middle Triassic time and included numerous species. *Cymbospondylus petrinus* is the species represented by the most complete fossil material in Nevada, but there may well have been other species in the Triassic seas of the Great Basin.

Cymbospondylus was an enormous predator, up to 30 feet long, but is considered a relatively primitive ichthyosaur. A long, straight tail propelled it with an undulating eel-like motion, but although the body was streamlined, it lacked a dorsal fin or tail flukes that are present in most advanced ichthyosaurs. The large, paddle-like forelimbs and hindlimbs were probably used to stabilize, maneuver, and brake the body as it prowled the seas. Most paleontologists view *Cymbospondylus* as a solitary cruiser of the open oceans, visiting the shallow, near-shore seas infrequently. The teeth of *Cymbospondylus* were sharply conical but generally less than an inch long, strikingly small for a predator with a body weight of 3–4 tons. The narrow jaw and dental characteristics suggest that the primary prey of *Cymbospondylus* was fish and cephalopods.

All the remains of *Cymbospondylus* in Nevada occur in calcareous shale or silty limestone associated with ammonoids and other molluscan fossils. These sediments were deposited in the relatively shallow water that covered central Nevada during most of the Triassic Period, but *Cymbospondylus* would certainly have been at home in the deeper open oceans of the western Great Basin.

As imposing as *Cymbospondylus* may have been, another gargantuan marine reptile was even more stunning and has become the quintessential icon of ichthyosaurs in western North America. In the 1890s, miners from the settlement of Berlin in the Shoshone Mountains of central Nevada, noticed strange disk-like fossils weathering from the rocks in Union Canyon. In 1928, after hearing of these fossils, Dr. Siemon Muller of Stanford University visited the area and correctly identified them as the remains of ichthyosaurs, and the disks were vertebrae. However, the amazing concentration of fossils in West Union Canyon did not become apparent until after 1954, when Drs. Charles Camp and Samuel Welles of the University of California, Berkeley, began large-scale excavations there.

FIGURE 8.15. Exposed portion of the *Shonisaurus* bone bed at the Berlin-Ichthyosaur State Park, central Nevada. Round objects in the lower foreground are individual vertebrae from the backbone, while long bones in the middle foreground are ribs.

Camp's work continued until the late 1960s, by which time he had discovered the remains of at least 37 different ichthyosaur individuals preserved in a thin interval of shaly limestone in the lower Luning Formation. At the time of their discovery, these fossils were regarded as belonging to the largest known ichthyosaur in the world. Camp named the new species *Shonisaurus popularis*, in reference to both the Shoshone Mountains where the remains were excavated and the number of individuals huddled at the site. Camp also identified two other species of *Shonisaurus*, but later paleontologists have debated their validity and their fossils were far less abundant than *S. popularis*. In 1957, the site of the excavation was protected by inclusion in the Berlin-Ichthyosaur State Park (BISP); a structure covering the exposed bone bed was constructed in 1966. The protected part of the fossil bed at the site is about 25 × 65 feet and contains the remains of nine large ichthyosaurs that have been left in situ. The fossil site was designated a National Natural Landmark by the U.S. National Park Service in 1973. Fittingly, *Shonisaurus popularis* was designated the State Fossil of Nevada in 1977 (Figure 8.15).

Shonisaurus, reaching an adult length of up to 50 feet and attaining a body mass as much as 40 tons, was the largest known resident of the Late Triassic seas of the western Great Basin. Comparable in size to the modern gray whale, it was considered the largest of all ichthyosaurs until fossils of even bigger relatives were discovered in Canada in 2004 and England in 2018. Nonetheless, this ichthyosaur would have been a stunning presence. Imagine such a giant reptile gliding through the water, propelling its deep, streamlined body with an undulating, muscular tail bearing a fin on its upper surface. Because it lacked large, paired tailfins, or flukes, *Shonisaurus* was probably not a very fast swimmer. In fact, *Shonisaurus* did not possess the large, slashing teeth of other ichthyosaurs but had smaller conical teeth restricted to the front of the lower and upper jaws. In adult specimens, the teeth were firmly and deeply set into sockets, but only 1–2 inches of the pointed crowns were exposed along the mouth. This suggests that *Shonisaurus* likely fed primarily on fish and squid-like cephalopods, both of which were plentiful in the Triassic seas. Some paleontologists have speculated that *Shonisaurus* may have lost its teeth as it grew larger and would then have ingested fish by simply swallowing them with a quick snap of its long, pointed snout. The mass mortality evident from the concentration of *Shonisaurus popularis* fossils at BISP suggests that this ichthyosaur was a gregarious species, roaming the seas in groups of dozens of individuals, like the pods of modern migratory whales.

The nature and origin of the bone bed in the Luning Formation at Berlin-Ichthyosaur State Park is intriguing and still somewhat mysterious. The preserved bones are nearly complete skeletons, with most of the bony elements in proper position relative to each other. Such articulated remains suggest minimal postmortem decomposition and transport of the carcasses. The shaly limestone matrix is unmistakably lithified carbonate mud that accumulated in water deep enough to escape strong wave or current agitation. Fossils of other swimming and floating animals, such as ammonites and conodonts, accompany the ichthyosaur remains, but remains of bottom-dwelling clams or crustaceans are rare. How did so many ichthyosaurs die and accumulate in a relatively small area on the shallow Triassic seafloor?

While causes of the mass mortality event remain uncertain, some paleontologists have suggested that the large reptiles might have died in a shallow bay, stranded after a sudden low tide left no means of escape to the open sea. However, the sediments containing the fossils appear to have accumulated under water too deep for such stranding to have been likely. Other scientists have suggested that the ichthyosaurs died singly in a shallow bay that was often visited for mating and/or birthing of young. Yet another hypothesis proposes that the bone bed originated in deeper water farther offshore when stagnant, poorly oxygenated bottom conditions developed seasonally. Perhaps, as was recently proposed, a harmful algal bloom, like a red tide caused by dinoflagellates in modern seas, triggered the death of an entire pod of these massive reptiles. Carcasses may have descended one at a time to a small area of stagnant, poorly oxygenated bottom water. In this scenario, normal postmortem decomposition would have been interrupted, allowing the carcasses to have remained intact long enough for burial under mud prior to disintegration.

These are all valid hypotheses, but none can be conclusively substantiated solely by geological evidence. Still, one other recent and well-publicized idea lacks any scientific evidence whatsoever: that a giant, squid-like monster (a "kraken") killed and placed the ichthyosaurs on the sea floor for later consumption. Not only is this idea absurdly fanciful and untestable, but paleontologists have no evidence whatsoever for the existence of such giant cephalopods in the Triassic seas.

The menagerie of Mesozoic marine reptiles in this period included one other enigmatic group, the thalattosaurs (order Thalattosauria, superfamily thalattosauroidea). Thalattosaurs are most common in Middle Triassic time and are only known from Europe (Switzerland, Italy, and Austria), eastern Asia, and western North America. These reptiles are closely related to the ichthyosaurs and have some of the same adaptations to marine conditions. Thalattosaurs generally had streamlined bodies, long tails, and fins or fleshy ridges along their backs. They probably swam using lateral undulations of the tail for propulsion. Most had pointed and sometimes downturned snouts with sharp, piercing teeth in the front of the jaw and more robust crushing and tearing teeth farther back (Figure 8.16). Thalattosaurs were usually medium-sized rep-

Skull of *Thalattosaurus alexandrae*

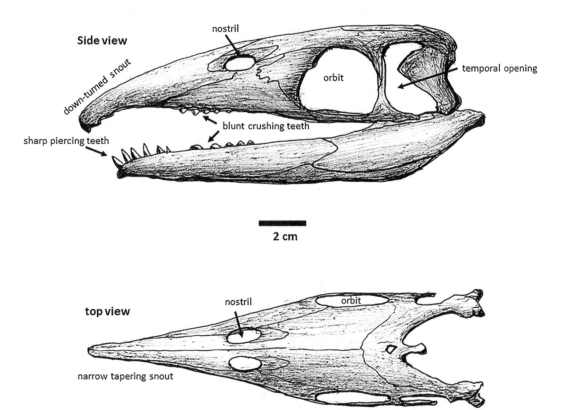

FIGURE 8.16. *Thalattosaurus alexandrae* skull in lateral view (*top*) and top view (*bottom*). This reconstruction is based on Triassic fossils from northern California but is probably like thalattosaurs that lived in the Great Basin seas.

tiles with a body length of about 5 feet and possessed relatively robust shoulder and limb bones with claws on their elongated feet. Some paleontologists have speculated that thalattosaurs might have lived in shallow coastal waters, waddling through the surf in search of prey. Unlike some of the giant ichthyosaurs that prowled the deeper offshore water, the thalattosaurs are typically found as isolated specimens, suggesting a more solitary mode of existence. Fragmentary thalattosaur remains have been found in the Late Triassic Natchez Pass Formation in the Buffalo Mountains southeast of Winnemucca, Nevada. The limestone and fine sandstones of the Natchez Pass Formation were deposited in shallow water during the final retreat of the seas

from the Great Basin. The Nevada fossils resemble the genera *Thalattosuarus* from northern California (Figure 8.16) and *Herscheleria* from Switzerland, indicating that the thalattosaurs ranged widely along the coastal margins of the Panthalassa Ocean.

The sauropterygian-ichthyosaur-thalattosaur assemblage of Triassic time represents a unique stage in the evolution of marine reptiles. Though some of these lineages would persist into the later periods of the Mesozoic Era, by about 100 million years ago, they had been mostly replaced by other, more advanced groups of seagoing reptiles such as the gigantic plesiosaurs, lizard-like mosasaurs, turtles, and sea crocodiles. The thalattosaurs became extinct near the end of the Triassic Period, and the ichthyosaurs died out in the Middle Cretaceous Period. We are unsure what caused this evolutionary turnover of marine reptiles, but the loss of habitat for those that were adapted to shallow water may have been a factor—at least in the Great Basin region. Recall that by the end of the Triassic Period, much of the Great Basin seafloor had been converted into dry land though the accretion of terranes to the western edge of Pangaea. Deeper oceans still existed farther west, where reptiles adapted to the open-ocean environment might have survived. When the interior of North America was flooded by the shallow seas in Late Mesozoic time, coastal marine habitats were reestablished along the interior seaway. By then, however, many of the Triassic forms had died out, leaving the mosasaurs, plesiosaurs, and crocodiles with little competition for the role of large reptilian predators in shallow coastal seas. Having dominated the oceans for 150 million years, the distinctive Triassic array of marine reptiles gave way to new mob of sea monsters. But this transition began as the seas withdrew from the Great Basin region for the final time.

Epilogue: The Great Stretch, Sinking Land, and Future Seas.

As we have seen, the sedimentary rocks of the Great Basin document more than 500 million years of seafloor conditions in what is now the arid deserts of western North America. The Great Basin seafloor was born with continental rifting around 700 million years ago in the Proterozoic Era. By Late Triassic time, about 200 million years ago, all but a few small areas of the seafloor had been converted into dry land. In the Jurassic Period, the seas completely withdrew from the Great Basin, except for the extreme western fringe of the region (Figure 8.17). After about 150 million years ago, the disappearance of the Great Basin seafloor was complete, and the seas never again returned to the region. At least, not yet.

The rock record of the seafloor tells the tale of a continually changing planet. Ancient con-

tinents such as Rodinia, Laurentia, Gondwana, and Laurussia bear little resemblance to North America—or any other modern landmass. Likewise, the oceans of past, Panthalassa and Mirovoi, were different from the modern seas in virtually every respect. On the dynamic Earth, patterns of land and sea are ceaselessly changing like the slow-motion turn of a kaleidoscope. In our quest to understand the deep history of life and land in the Great Basin, we have relied on the ingenious methods developed by scientists for teasing clues about past conditions from rocks and fossils. However, as certainly as the earth has a past, it also has a future. Our planet's plate tectonic system is still active, and the evolution of the physical earth—and the life it sustains—will extend far beyond our own present experience. Given the 700-million-year pageant of continually shifting land and ocean, can we really be certain that the seas of the Great Basin are gone forever? Of course not.

After the Jurassic withdrawal of seas from the Great Basin region, convergent plate interactions along the western margin of North America continued lifting the land, igniting volcanic activity across the region. From Late Mesozoic time into the middle part of the Cenozoic Era, compressive deformation buckled the crust, while eruptions piled thick layers of lava and ash across the central Great Basin. The land rose some 9,000 feet above sea level. Then, about 30 million years ago, as portions of a midocean ridge were subducted eastward under North America, the plate interactions along the western margin of the continent began shifting from a convergent type to the modern transform boundary between the Pacific and North American plates. This transition initiated the San Andreas fault system along the California coast and replaced the compression related to the earlier plate convergence with westward extension and northwest shearing. In response to this rearrangement, the elevated land in the Great Basin began collapsing, and the underlying crust was stretched west and northwest relative to the interior of North America. Today, the Great Basin region is roughly twice as wide and

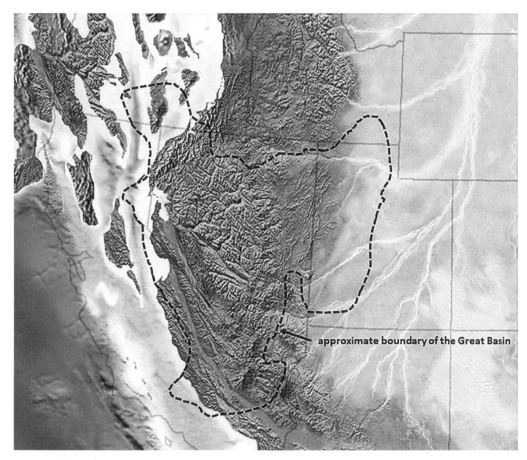

approximate boundary of the Great Basin

FIGURE 8.17. Paleogeography of western North America in late Jurassic time, 150 million years ago. Paleogeographic base map prepared by R. Blakey, Colorado Plateau Geosystems Inc.

nearly a mile lower in elevation than in Early Cenozoic time.

The extension and collapse of the Great Basin crust that began in Middle Cenozoic time is still underway. Modern methods of measuring plate motion, such as satellite radar interferometry and precise GPS (Global Positioning System) mapping, have revealed that the western edge of the North American plate is still stretching westward, and the Basin is sinking as it does so. The forces driving this deformation are generated along the transform boundary between the North American plate and the Pacific plate. Plate motion vectors (Figure 8.18) indicate that the Pacific plate, the largest on the Earth, is moving rapidly to the northwest relative to the stable interior of the North American plate. The lateral motion is slightly oblique and generates a force that geologists call "transtension," a term combining the transform plate interactions and the tensional stress generated inward from it. In simple terms, as the vast, oceanic plate slides northwest, it pulls away a bit from North America, stretching the continental crust behind it. North of the Mendocino coast in California, the small Gorda plate is converging against the edge of North America, generating compression associated with the Cascadia subduction zone. The three plates, Pacific, North American, and Gorda, meet at a point offshore of the coast of northern California known as the Mendocino triple junction (Figure 8.18). The northern limit

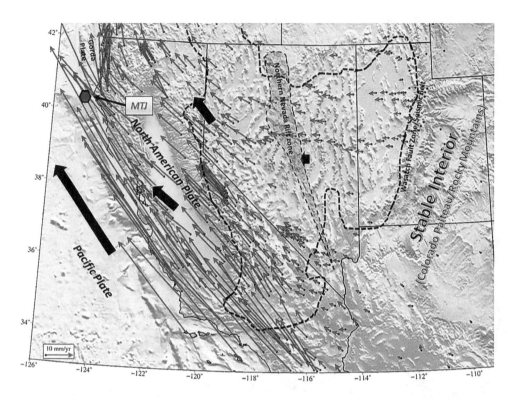

FIGURE 8.18. GPS plate motion vectors for the Great Basin region: small red arrows denote vectors from individual stations, while larger black arrows depict overall motions of portions of the North American and Pacific plates. MTJ = Mendocino Triple Junction. Modified from Nevada Bureau of Mines and Geology Map M178, A Geodetic Strain Rate Model for the Pacific-North American Plate Boundary (2012).

of the Great Basin closely coincides with the eastward projection of triple junction, an indication of the relationship between the transform plate interactions and the tensional forces ripping the Great Basin.

The stretching forces that reach inland from the Pacific–North American plate boundary are distending the crust beneath the Great Basin westward 2–3 mm/year across most of the region. This rate of the deformation increases westward to more than 10 mm/year, and the direction shifts to the northwest along the western edge of the Great Basin. The east–west increase in deformation and shift in direction is a direct consequence of the immense forces generated by the fast-moving Pacific plate grinding the continent along the San Andreas system and other related fault zones in California and west-

ern Nevada. These transtensional stresses have broken the crust into fault blocks, creating the basin-and-range topography of the Great Basin and facilitating the continuing collapse of the former highland. The modern plate tectonic interactions will persist for many millions of years, and tensional forces will continue to pull the crust of North America unevenly west and northwest. The land will continue to collapse and, ultimately, portions of the crust will again fall below sea level.

Geophysical studies and seismic patterns across the Great Basin indicate that there are several linear zones in the region that are deforming at higher-than-average rates. These zones, in part reflecting older, deeply buried structures, may be the first areas to drop below sea level. At current deformation rates, by some

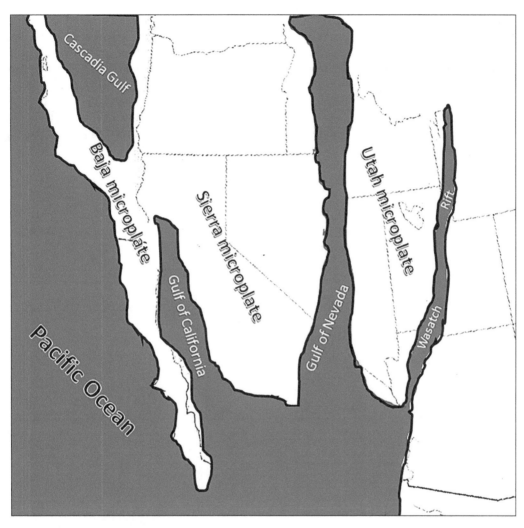

FIGURE 8.19. A plausible future geography of the Great Basin region, 70 million years from now.

70 million years in the future these rift zones will likely form elongated embayments that herald the return of oceanic conditions to the Great Basin (Figure 8.19). Between the submerged rift zones, the crust of western North America will be broken into several microplates. Subsequent geography will be shaped by their movement. Because the western edge of the North American plate will disintegrate into a complex pattern of microplates, it is difficult to project geography beyond the initial return of seas into the Great Basin region. In any case, the obvious collapse and stretching of the Great Basin over the recent geologic past leaves us with only one certainty about the future: the Great Basin seafloor, high and dry for the time being, will be submerged.

Humanity, at least in the form with which we are familiar, will not witness these events. But our evolutionary descendants may someday ponder the world we know today and attempt its reconstruction just as we have utilized geological evidence to investigate the Great Basin seafloor of our past. The modern Great Basin landscape will be represented in the future rock record as a widespread unconformity produced by the erosion of bedrock in modern mountain ranges. This unconformity will be associated with thick, but discontinuous, masses of conglomerate, sandstone, and shale formed from the sediments currently accumulating in alluvial fans, dry washes, and ephemeral lakes. Fossils in these sediments may include bones of coyotes, badgers, rattlesnakes, antelope, deer, rodents, and rabbits. Perhaps the sediments deposited around springs will have fossilized leaves of rushes and willows. Geologists of the future may also notice crushed beer cans, ancient car parts, pockets of radioactive isotopes, decaying barrels of toxic waste, and bits of electronic refuse as well. What visions will they form from this evidence of the Great Basin landscape we know so well today? What perceptions of *us* will be fostered by the evidence we leave behind in the rock record of the future?

Glossary

Abyssal: pertaining to ocean depths exceeding 3,500 feet.

Allochthon: a mass of rock, typically deformed internally, that has been tectonically transported from its place of origin. In the Great Basin, allochthons are commonly transported along underlying thrust faults.

Anoxic: an environment without oxygen.

Archaea: one of the three domains of life consisting of very small, primitive (prokaryotic), unicellular organisms resembling bacteria but with different genetic and biochemical makeup.

Archaeocyathids: organisms belonging to the taxonomic group Archaeocyatha, regarded as closely related to the sponges in phylum Porifera or perhaps a separate, but similar phylum. The archaeocyathids typically have double, cone-in-cone, perforated walls and may have a conical or branching shape. Archaeocyathids are restricted to Cambrian-age rocks.

Aragonite: carbonate mineral identical in composition to calcite ($CaCO_3$) but with different crystal symmetry.

Argillite: a compact, fine-grained, clay-rich sedimentary rock that is typically more thoroughly lithified than ordinary siltstone or shale. Most argillite lacks laminations and has the appearance of a solidified clay paste.

Asthenosphere: a zone of partially molten, semi-rigid rock in the upper mantle immediately below the lithosphere. The top of the asthenosphere is about 100 km beneath the Earth's surface, and the base is between 350 and about 700 km deep.

Autotropic: any organism that manufactures its own nutrients from inorganic components.

Benthic: anything on the bottom of a body of water, generally on the sediment surface or slightly below.

Bolide: an extraterrestrial object such as an asteroid, meteoroid, or comet that either explodes in the atmosphere or strikes the surface with enough force to excavate a crater and generate shock waves.

Brachiopod: a bivalved suspension-feeding invertebrate belonging to the phylum Brachiopoda. Brachiopods are superficially like clams of the phylum Mollusca, except that the plane of symmetry passes through, not between, the two shells. The internal anatomy of brachiopods is also quite distinct from any mollusc in (among other things) the presence of a *lophophore*, a stiff structure that supports the gills and feeding filaments.

Bryozoan: invertebrate organisms belonging to the phylum Bryozoa, consisting chiefly of colonial animals that construct calcareous exoskeletons.

Bioclastic: a textural term applied to sedimentary rocks that contain abundant fragmental material of organic origin such as shell fragments, fecal pellets, or skeletal grains.

Calcareous: a term indicating that a substance is composed mostly of calcium carbonate ($CaCO_3$)—in any mineral form.

Calcite: a carbonate mineral of the composition $CaCO_3$.

Carbonate: any mineral composed of carbonate ions (CO_3^{-2}) combined with other metals or nonmetals. Calcite ($CaCO_3$), dolomite ($CaMa(CO_3)_2$), and aragonite ($CaCO_3$) are the most common carbonate minerals in sedimentary rocks.

Carpoids: early members of the phylum Echinodermata.

Chert: a hard and compact chemical sedimentary rock consisting of microcrystalline quartz (crystals smaller than 30 micrometers). Flint is a common term for dark-colored varieties of chert, while reddish-brown chert is known as jasper.

Comet: a relatively small celestial body composed of frozen water and gases, rock, and dust particles that typically orbit the sun beyond the major planets. Comets can be perturbed from their normal orbits to pass through the inner solar system where collisions with planets are possible.

Coquina: a variety of limestone consisting mainly of coarse shell fragments and shelly debris.

Craton: the stable geological core of a continent, generally consisting of Precambrian rocks that have not been more recently deformed.

Crinoid: an invertebrate of the class Crinoidea, phylum Echinodermata. Known commonly as "sea lilies," the exoskeleton of crinoids was comprised of an inverted cup consisting of calcareous plates supported on a stem of circular plates (ossicles). Multiple arms extended from the cup aperture into the water column to capture food from passing currents.

Cyanobacteria: formerly called blue-green algae, the cyanobacteria are mostly aerobic photosynthetic microbes that can form colonies and surface mats in shallow marine environments. Cyanobacteria are thought to have been primary builders of stromatolites and thrombolites in Proterozoic and Paleozoic sedimentary rocks.

Detrital: a term describing granular material—or sedimentary rocks comprised of it—such as sand grains. Detrital particles are produced by the physical disintegration of rock as part of the weathering process.

Diagenesis: a broad term encompassing all chemical, physical, or biological change that occurs in sediment after its initial deposition or precipitation. The adjective **diagenetic** refers to any attribute of a sedimentary rock that develops during diagenesis.

Diamictite: a textural term describing a sedimentary rock composed of a wide range of particle sizes regardless of the origin of sediment. Diamictites are generally like conglomerates but commonly differ from them by the presence of very fine sand, silt, or clay particles along with the larger components.

Dolomite: a carbonate mineral consisting of calcium magnesium carbonate, $CaMg(CO_3)_2$. The term is commonly applied to both the mineral and a sedimentary rock consisting of it. Alternatively, the term "dolostone" is used to distinguish the rock from the mineral.

Dysoxic: a term describing an environment of low oxygen levels.

Echinoderm: any member of the phylum Echinodermata, a large group of marine invertebrates with radial symmetry. Included within the Echinodermata are the living sea stars, sand dollars, sea urchins, crinoids (sea lilies), and brittle stars, along with many extinct groups.

Eocrinoids: an extinct group of stalked echinoderms crudely similar to the more advanced crinoids. Eocrinoids were abundant in the Cambrian Period but became extinct in Silurian time.

Eugeocline: an older term used to describe deep marine sediments that accumulate seaward of a shallow continental shelf (miogeocline). Shale, fine-grained sandstone, and chert that formed on the continental slope and abyssal seafloor are typical of eugeoclinal strata in the Great Basin.

Eustatic: pertaining to worldwide changes in sea level caused by global changes in climate, plate tectonic events, ocean basin evolution, or some combination of these factors.

Evaporite: chemical sedimentary rocks that consist of minerals precipitated from saline water because of evaporation. Common evaporite minerals include halite ("rock salt," sodium chloride), gypsum or anhydrite (both calcium sulfate minerals), and calcite (calcium carbonate).

Foreland: an area bordering an orogenic belt adjacent to a rising highland. Foreland regions usually separate an active orogenic highland from the stable portion of a continental margin.

Foredeep: an elongated basin formed by rapid subsidence along the margin of an active orogenic belt.

Ga: Giga-annum; one billion years before the present.

Glaciomarine: any process, interaction, or

consequence of glacial ice in an oceanic environment.

Graded bedding: a sedimentary deposit exhibiting vertical changes in grain size, usually from coarse-grained (granules, pebbles, etc.) at the bottom to fine-grained (silt, sand, clay) at the top. Graded bedding is commonly the result of deposition from undersea turbidity currents.

Intertidal: pertaining to that portion of the seafloor between normal high tide and low tide. The intertidal zone (also known as the littoral zone) is completely submerged at high tide and exposed during low tide.

Ion: an atom or molecule that bears an electrical charge due to unequal numbers of positively charged protons and negatively charged electrons.

Laurentia: an ancient continent, predecessor of modern North America, that existed between the Late Proterozoic rifting of the supercontinent Rodinia and the assembly of supercontinent Pangaea in early Mesozoic time. Laurentia included rocks comprising the core of modern North America and Greenland, along with all material deposited or accreted along the margins prior to its amalgamation into Pangaea.

Limestone: a chemical sedimentary rock composed dominantly of the mineral calcite precipitated directly from sea water as tiny crystals or through the activity of organisms such as calcareous algae, coral, or molluscs.

L chondrites: meteorites with relatively low iron abundance. They are thought to have formed from the destruction of the same parent body in the asteroid belt.

Lithification: a general term for the slow process of compacting and cementing soft sediment into solid sedimentary rocks.

Lithosphere: the outer rocky shell of the earth, averaging 100 km thick, composed of either oceanic or continental crust (or both) overlying dense rock of the uppermost mantle. The modern lithosphere is broken into about 20 mobile plates.

Lycopods: an extinct group of primitive trees known as "scale trees" due to the diamond-shaped leaf-scars that spiral around the trunk. Paleobotanists assign these ancient trees to various plant categories including the Lycophyta, Lycopsida, Lycopodiophytina, and others, creating a somewhat confusing tangle of taxonomic names for these prehistoric trees.

Karst: surface topography that results from the dissolution of soluble rock such as limestone, marble, or rock salt by fresh water on the surface or below ground. Karst topography is characterized by caverns, sinkholes, solution valleys, solution fissures, and underground drainage.

Karstification: the process of developing Karst topography that results from the exposure of rocks comprised of soluble minerals such as calcite, dolomite, halite, or gypsum to surface conditions.

Ma: Mega-annum, one million years before present.

Magma: molten rock that originates underground within the lithosphere, as opposed to lava, the term applied to magma on the earth's surface.

Micrite: a fine-grained variety of limestone consisting of microscopic calcite crystals smaller than 4 micrometers (10^{-6}). Micrite is essentially lithified carbonate mud.

Microbialite: an organo-sedimentary rock, commonly a laminated carbonate, comprised of sediment deposited under the influence of microbial mats on the seafloor. Microbialite deposits accumulate most readily in shallow, warm, clear seas. Structures preserved in microbialites may include stromatolites, wavy lamination, and thrombolites.

Miogeocline: an area of shallow marine sediment accumulation along a passive continental margin. Sediment deposited on the miogeocline thickens seaward from the shoreline and typically grades into deep-water deposits such as shale, argillite, and chert farther offshore.

Nektonic: a term describing any organism capable of free swimming and migration through bodies of water.

Nonconformity: a specific type of unconformity that separates crystalline igneous or metamorphic rocks from overlying sedimentary strata.

Because the underlying crystalline rocks form at depth within the crust, nonconformities represent gaps in the rock record corresponding to the time required for uplift and erosion prior to deposition of sediment grains on the erosional surface.

Olistolith: a large block of exotic rock transported by undersea sliding and gravity flow into a depositional basin. Olistoliths are commonly found in masses of gravity-transported sediment known as olistostromes.

Olistostrome: a chaotic deposit of mixed rocky rubble transported as a debris-flow into a basin of accumulation.

Oncolites: rounded masses of microbially trapped sediment grains, generally with concentric laminations. Oncolites vary in size, rarely exceeding 5 cm (2 inches) in diameter and are typically spherical.

Orogeny: a traditional term describing a mountain-building event. Orogenies result from plate tectonic interactions that cause an uplift of mountains from crustal thickening by deformation and magmatism.

Orogenic belt: a linear region of deformation, magmatism, and metamorphism most related to convergent plate tectonic interactions. Orogenic belts eventually rise to form generally linear mountain systems underlain by highly deformed rock.

Ostracoderm: a general term used to describe any of six classes of jawless armored fish of the Paleozoic Era. All ostracoderms are distantly related to modern lampreys and hagfish.

Paleokarst: ancient karst surface features and subsurface cavity in-fillings preserved in sedimentary rock after deposition of younger sediment over previously exposed strata.

Paleosol: an ancient soil horizon preserved with a stratigraphic succession. Paleosols form when rocks are exposed to weathering for periods of time long enough for soil profiles to develop from bedrock.

Panthalassa: the single large ocean (or "super-ocean") that surrounded the supercontinent of Pangaea when it formed about 250 million years ago.

Phosphorite: any sedimentary rock, typically a carbonate, that contains high concentrations of phosphorous in the form of phosphate minerals such as apatite group, $Ca_5(PO_4)_3$ (F,Cl,OH).

Photoautotroph: any organism—unicellular or multicellular—that uses light energy to synthesize its own nutrients.

Phyllocarid: a crustacean arthropod belonging to the subclass Phyllocarida, possessing a two-part laterally flattened shell, long tail, and jointed legs.

Placoderm: any member of the class Placodermi, an extinct group of Paleozoic armored fish. The placoderms are a diverse class with at least eight different orders recognized by paleontologists.

Planktonic: a term describing any organism with the ability to float, or feebly swim in the water column while passively drifting with currents.

Reflux: the return flow of water from a secondary basin to the reservoir of its origin; a term used to describe the seepage of hypersaline brines from an evaporite basin to the seafloor.

Rodinia: an ancient supercontinent that existed during the Neoproterozoic Era, from about 1 billion Ga to about 650 Ma.

Sabkha: a salt-encrusted, near-shore supratidal basin periodically flooded by sea water at high tide. Evaporation of water isolated from the open ocean creates dense hypersaline brine resulting in evaporite deposits on the floor of the sabkha.

Sessile: any plant or animal resting on, or attached to, the seafloor with no capacity for active motion.

Shale: a thin-bedded sedimentary rock predominantly composed of clay-sized (<1/256 mm) particles.

Siliciclastic: sediment or sedimentary rock consisting primarily of quartz grains (of any size) or other silicate minerals.

Stromatolite: a laminated sedimentary structure that results from sediment trapping, mineral precipitation, and particle binding association with the growth of microbial colonies on the seafloor or a lake bottom.

Stromatoporoids: a group problematic extinct organisms currently thought to be calcareous sponges in the phylum Porifera. Stromato-poroids constructed calcareous skeletons of many forms, including hemispherical, bulbous, branching, and encrusting.

Supratidal: pertaining to the near-shore portion of a coastline above normal high-tide level.

Subtidal: the shallow ocean zone below normal low tide to the edge of the continental shelf. Under normal conditions, the subtidal zone is permanently submerged.

Terrane: a discrete mass of oceanic or continental material added to the core, or craton, of a continental block by accretion related to plate convergence. Terranes generally consist of deformed and fault-bounded blocks of rock dissimilar to the material that surrounds them.

Thrombolite: a clotted, nonlaminated sedimentary structure formed from trapping and cementation of sedimentary grains by biofilms of microorganisms, especially cyanobacteria. Thrombolites may be branching, nodular, or irregular.

Thrust, thrust fault: a low-angle fault across which the block of rock overlying the fault plane is displaced upward relative to the block beneath the fault plane. Thrust faults only form from compressive forces.

Till: poorly sorted, generally unstratified sediment deposited directly from melting glacial ice.

Tillite: a sedimentary rock composed of lithified till.

Trace fossils: biogenic sedimentary structures formed by the activity of living organisms and preserved in sedimentary rocks. Common trace fossils include tracks, trails, burrows, and footprints created by organisms moving over or through soft sediment.

Transpression: stresses that result from the combination of compression generated by plate convergence and lateral shear related to sliding movements along a transform plate boundary. Transpression is produced during oblique convergence of plates.

Trilobite: an organism belonging to a group (class Trilobita) of extinct arthropods characterized by a segmented body divided into three longitudinal lobes. Trilobites evolved in the early Cambrian Period and became extinct at the end of Permian time.

Turbidite: a granular sedimentary rock, typically with graded bedding and scour marks, formed from sediment grains transported by a turbidity current.

Turbidity current: a dense current of water mixed with rock particles flowing rapidly along the ocean bottom down a steep submarine slope. Turbidity currents are essentially undersea "landslides" that can instantly deposit great amounts of sediment in a fan-shaped mass in deep-ocean basins. Turbidity currents can be generated along continental margins by earthquakes, strong storms, and violent volcanic eruptions.

Weathering: the spontaneous process of decay that affects rocks exposed at the Earth's surface. Weathering results from several different chemical reactions and physical processes that degrade or weaken the mineral grains in bedrock.

References

Chapter 1. Vanished Oceans of the Great Basin Desert

DeCourten, F. L. 1993. *The Broken Land: Adventures in Great Basin Geology*. Salt Lake City: University of Utah Press.

DeCourten, F., and N. Biggar. 2017. *Roadside Geology of Nevada*: Missoula, MT: Mountain Press.

Dickinson, W. R. 2013. "Phanerozoic Palinspastic Reconstructions of Great Basin Geotectonics (Nevada-Utah, USA)." *Geosphere* 9, no. 5: 1384–1396.

Fiero, B. 2009. *Geology of the Great Basin*. Reno: University of Nevada Press, Reno.

Francaviglia, Richard V. 2005. *Mapping and Imagination in the Great Basin: A Cartographic History*. Reno: University of Nevada Press.

Hintze, L. F., and F. D. Davis. 2003. "Geology of Millard County, Utah." *Utah Geological Survey Bulletin* 133.

Inkenbrandt, C. 2016. "Contributions to the Understanding of Geology in Utah's West Desert." In *Resources and Geology of Utah's West Desert*, edited by J. B. Comer, C. Inkenbrandt, K. A. Krahulec, and M. L. Pinnell. Utah Geological Association Publication 45:1–12.

Simpson, J. H. 1876. *Report of Exploration across the Great Basin of the Territory of Utah for a Direct Wagon Route from Camp Floyd to Genoa, Carson Valley, in 1859*. Washington, D.C., Government Printing Office

Stewart, J. H. 1980. *Geology of Nevada: A Discussion to Accompany the Geologic Map of Nevada*. Special Publication 4. Reno: Nevada Bureau of Mines and Geology, University of Nevada.

Trimble, S. 1989. *The Sagebrush Ocean: A Natural History of the Great Basin*. Reno: University of Nevada Press.

Walker, J. D., J. W. Geissman, S. A. Bowring, and L. E. Babcock, compilers. 2018. Geologic Time Scale v. 5.0. Boulder, CO: Geological Society of America.

Chapter 2. The Oceans Arrive

Anbar, A. D., and A. D., Knoll. 2002. "Proterozoic Ocean Chemistry and Evolution: A Bioinorganic Bridge?" *Science* 297:1137–1142.

Bogdanova, S. V., S. A. Pisarevsky, and X. Z. Li. 2009. "Assembly and Breakup of Rodinia: Some Results of IGCP Project 440." *Stratigraphy and Geological Correlation* 17, no. 3: 259–274.

Bond, G. C., N. Christie-Blick, M. A. Kominz, and W. J. Devlin. 1985. "An Early Cambrian Rift to Post-Rift Transition in the Cordillera of Western North America." *Nature* 316:742–745.

Cai, Y., S. Xiao, G. Li, and H. Hua. 2019. "Diverse Biomineralizing Animals in the Terminal Ediacaran Period Herald the Cambrian Explosion." *Geology* 47:380–384.

Corsetti, F. A., S. M. Awramik, and D. Pierce. 2003. "A Complex Microbiota from Snowball Earth Times: Microfossils from the Neoproterozoic Kingston Peak Formation, Death Valley, USA." *Proceedings of the National Academy of Sciences* 100(8):4399–4404.

Corsetti, F. A., and A. J. Kaufman. 2003. "Stratigraphic Investigations of Carbon Isotope Anomalies and Neoproterozoic Ice Ages in Death Valley, California." *Geological Society of America Bulletin* 115:916–932.

Creveling, J. R., K. D. Bergman, and J. P. Grotzinger. 2016. "Cap Carbonate Platform Facies

Model, Noonday Formation, SE California." *Geological Society of America Bulletin* 128: 1249–1269.

Darroch, S. A. F., E. F. Smith, M. Laflamme, and D. H. Erwin. 2018. "Ediacaran Extinction and Cambrian Explosion." *Trends in Ecology and Evolution* 33, no. 9:653–663.

Dehler, C. M., D. A. Sprinkel, and S. M. Porter. 2005. "Neoproterozoic Uinta Mountain Group of Northeastern Utah: Pre-Sturtian Geographic, Tectonic, and Biologic Evolution." In *Geological Society of America Field Guide 6: Interior Western United States*, edited by J. Pederson and C. M. Dehler, 1–25. Boulder, CO: Geological Society of America.

Dehler, C. M., G. Gehrels, S. Porter, M. Heizler, K. Karlstrom, G. Cox, L. Crossey, and M. Timmons. 2017. "Synthesis of the 780–740 Ma Chuar, Uinta Mountain, and Pahrump (ChUMP) Groups, Western USA: Implications for Laurentia-Wide Cratonic Marine Basins." *Geological Society of America Bulletin* 129, no. 5–6:607–624.

Halverson, G. P., B. P. Wade, M. T. Hurtgen, and K. K. Barovich. 2010. "Neoproterozoic Chemostratigraphy." *Precambrian Research* 182, issue 4:337–350.

Harwood, C. L., and D. Y. Sumner. 2011. "Microbialites of the Neoproterozoic Beck Spring Dolomite, Southern California." *Sedimentology* 58, no. 6:1365–1391.

———. 2012. "Origins of Microstructures in the Neoproterozoic Beck Spring Dolomite: Variations in Microbial Community and Timing of Lithification." *Journal of Sedimentary Research* 82, no. 9:709–722.

Horodyski, R. J., and L. P. Knauth. 1994. "Life on Land in the Precambrian." *Science* 263, no. 5146:494–498.

Knauth, L. P., and M. J. Kennedy. 2009. "The Late Precambrian Greening of the Earth." *Nature* 460, no. 7256:728–732.

Le Heron, D. P., M. E. Busfield, and A. R. Prave. 2014. "Neoproterozoic Ice Sheets and Olistoliths: Multiple Glacial Cycles in the Kingston Peak Formation, California." *Journal of the Geological Society* 171:525–538.

Li, Z. X., et al. 2008. "Assembly, Configuration, and Break-Up of Rodinia: A Synthesis." *Precambrian Research* 160, issues 1–2:179–210.

Li, Z. X., D. A. D. Evans, and G. P. Halverson. 2013. "Neoproterozoic Glaciations in a Revised Global Palaeogeography from the Breakup of Rodinia to the Assembly of Gondwanaland." *Sedimentary Geology* 294:219–232.

Link, K., and N. Christie-Blick. 2011. "Neoproterozoic Strata of Southeastern Idaho and Utah: Record of Cryogenian Rifting and Glaciation." *Geological Society of London Memoirs* 36:425–436.

Macdonald, F. A., A. R. Prave, R. Petterson, E. F. Smith, S. B. Pruss, K. Oates, F. Waechter, D. Trotzuk, and A. E. Fallick. 2013. "The Laurentian Record of Neoproterozoic Glaciation, Tectonism, and Eukaryotic Evolution in Death Valley, California." *Geological Society of America Bulletin* 125, no. 7–8:1203–1223.

Mahon, R. C., C. M. Dehler, P. K. Link, K. E. Karlstrom, and G. E. Gehrels. 2014. "Geochronologic and Stratigraphic Constraints on the Mesoproterozoic and Neoproterozoic Pahrump Group, Death Valley, California: A Record of the Assembly, Stability, and Breakup of Rodinia." *Geological Society of America Bulletin* 126, nos. 5–6:652–664.

Maxwell, A. L., M. W. Wallace, A. van Smeerdijk, A. Hood, and N. Planavsky. 2018. "Cryogenian Iron Formations in the Glaciogenic Kingston Peak Formation, California." *Precambrian Research* 310:443–462.

McBride, E. F. 2012. "Petrology of the Eureka Quartzite (Middle and Late Ordovician), Utah and Nevada, U.S.A." *Rocky Mountain Geology* 47, no. 2:81–111.

Miller, J. M. G. 1985. "Glacial and Syntectonic Sedimentation: The Upper Proterozoic Kingston Peak Formation, Southern Panamint Range, Eastern California." *Geological Society of America Bulletin* 96:1537–1553.

Mrofka, D. D. 2010. "Competing Models for the Timing of Cryogenian Glaciation: Evidence from the Kingston Peak Formation, Southeastern California." PhD diss., University of California, Riverside. ProQuest (Mrofka_ucr_0032D_10206), https://escholarship.org/uc/item/3ph3f3ps.

Mrofka, D. D., and M. Kennedy. 2011. "The Kingston Peak Formation in the Eastern Death Valley Region." In *The Geological Record of Neoproterozoic Glaciations*, edited by E. Arnaud, G. P. Halverson, and G. Shields-Zhou, 449– 458. Geological Society of London Memoirs 36.

Petterson, R., A. R. Prave, and B. P. Wernicke. 2011. "Glaciogenic and Related Strata of the Neoproterozoic Kingston Peak Formation in the Panamint Range, Death Valley Region, California." In *The Geological Record of Neoproterozoic Glaciations*, edited by E. Arnaud, G. P. Halverson, and G. Shields-Zhou, 459–465. Geological Society of London Memoirs 36.

Schiffbauer, J. D., J. W. Huntley, G. R. O'Neil, S. A. F. Darroch, M. Laflamme, and Y. Cai. 2016. "The Latest Ediacaran Wormworld Fauna: Setting the Ecological Stage for the Cambrian Explosion." *GSA Today* 26, no. 11: 4–11.

Schirber, M. 2015. "Snowball Earth Might Be Slushy." *Astrobiology at NASA: Life in the Universe*, August 27, 2015. U.S. National Aeronautics and Space Administration. https://astrobiology.nasa.gov/news/snowball-earth-might-be-slushy/.

Shuster, A. M., M. W. Wallace, A. v. S. Hood, and G. Jiang. 2018. "The Tonian Beck Spring Dolomite: Marine Dolomitization in a Shallow, Anoxic Sea." *Sedimentary Geology* 368:83–104.

Smith, E. F., F. A. Macdonald, J. L. Crowley, E. B. Hodgin, and D. P. Schrag. 2015. "Tectonostratigraphic Evolution of the *c*. 780–730 Ma Beck Spring Dolomite: Basin Formation in the Core of Rodinia." *Geological Society, London, Special Publications*, 424:213–239.

Smith, E. F., L. L. Nelson, M. A. Strange, A. E. Eyster, S. M. Rowland, D. P. Schrag, and F. A. Macdonald. 2016. "The End of the Ediacaran: Two New Exceptionally Preserved Body Fossils Assemblages from Mount Dunfee, Nevada." *Geology* 44, no. 11:911–914.

Smith E. F., L. L. Nelson, S. M. Tweedt, H. Zeng, and J. B. Workman. 2017. "A Cosmopolitan Late Ediacaran Biotic Assemblage: New Fossils from Nevada and Namibia Support a Global Biostratigraphic Link." *Proceedings of the Royal Society B* 284:20170934.

Vandyke, T. M., D. P. Le Heron, D. M. Chew, J. M. Amato, M. Thrilwall, C. M. Dehler, J. Hennig, et al. 2018. "Precambrian Olistoliths Masquerading as Sills from Death Valley, California." *Journal of the Royal Society* 175(3): 377–395.

Witkosky, R., and B. P. Wernicke. 2018. "Subsidence History of the Ediacaran Johnnie Formation and Related Strata of Southwest Laurentia: Implications for the Age and Duration of the Shuram Isotopic Excursion and Animal Evolution." *Geosphere* 14, no. 5: 2245–2276.

Chapter 3. The Great Cambrian Explosion

Briggs, D. E. G., S. Halgedahl, and R. D. Jarrad. 2005. "A New Metazoan from the Middle Cambrian of Utah and the Nature of the Vetulicolia." *Palaeontology* 48, no. 4:681–686.

Briggs, D. E. G., B. S. Lieberman, J. R. Hendricks, S. L. Halgedahl, and R. D. Jarrard. 2008. "Middle Cambrian Arthropods from Utah." *Journal of Paleontology* 82, no. 2:238–254.

Cordie, D. R., S. Q. Dornbos, and P. J. Marenco. 2019. "Increase in Carbonate Contribution from Framework-Building Metazoans through Early Cambrian Reefs of the Western Basin and Range, USA." *Palaios* 34, no. 3:159–174.

Corsetti, F. A. and J. W. Hagadorn. 2003. "Precambrian–Cambrian Transition in the Southern Great Basin, USA." *The Sedimentary Record* 1, no. 1:4–8.

Elrick, M., S. Rieboldt, M. Saltzman, and R. M. McKay. 2011. "Oxygen-Isotope Trends and Seawater Temperature Changes across the Late Cambrian Steptoean Positive Carbon-Isotope Excursion (SPICE Event)." *Geology* 39, no. 10:987–990.

English, A. M., and L. E. Babcock. 2010. "Census of the Indian Springs Lagerstätte, Poleta Formation (Cambrian), Western Nevada, USA." *Palaeogeography, Palaeoclimatology, Palaeoecology* 295:236–244.

Foster, J. 2014. *Cambrian Ocean World: Ancient Sea Life of North America*. Bloomington: Indiana University Press.

Gaines, R. R., and M. L. Droser. 2005. "New Approaches to Understanding the Mechanics of Burgess Shale-Type Preservation: From Micron Scale to Global Picture." *The Sedimentary Record* 3, no. 2:4–8.

Gaines, R. R., M. J. Kennedy, and M. L. Droser. 2005. "A New Hypothesis for Organic Preservation of Burgess Shale Taxa in the Middle Cambrian Wheeler Formation, House Range, Utah." *Palaeogeography, Palaeoclimatology, Palaeoecology* 220:193–205.

Gill, B. C., T. W. Lyons, S. A. Young, L. R. Kunp, A. H. Knoll, and M. R. Saltzman. 2011. "Geochemical Evidence for Widespread Euxinia in the Later Cambrian Ocean." *Nature* 469: 80–83.

Hollingsworth, J. S. 2005. "A Trilobite Fauna in a Storm Bed in the Poleta Formation (Dyeran, Lower Cambrian), Western Nevada, USA." *Geosciences Journal* 9, no. 2:129–143.

Karlstrom, K., J. Hagadorn, G. Gehrels, W. Matthews, M. Schmitz, L. Madronich, J. Mulder, M. Pecha, D. Giesler, and L. Crossey. 2018. "Cambrian Sauk Transgression in the Grand Canyon Region Redefined by Detrital Zircons." *Nature Geoscience* 11:438–443.

Kepper, J. C. 1981. "Sedimentology of a Middle Cambrian Outer Shelf Margin with Evidence for Syndepositional Faulting, Eastern California and Western Nevada." *Journal of Sedimentary Petrology* 51:807–821.

Lieberman, B. 2003. "A New Soft-Bodied Fauna: The Pioche Formation of Nevada." *Journal of Paleontology* 77, no. 4:674–690.

Linnemann, U., M. Ovtcharova, U. Schaltegger, and A. Gartner. 2019. "New High-Resolution Age Data from the Ediacaran-Cambrian Boundary Indicate Rapid, Ecologically Driven Onset of the Cambrian Explosion." *Terra Nova* 31:49–58.

Mata, S. A., C. L. Corsetti, F. A. Corsetti, S. M. Awramik, and D. J. Bottjer. 2012. "Lower Cambrian Anemone Burrows from the Upper Member of the Wood Canyon Formation, Death Valley Region, United States: Paleoecological and Paleoenvironmental Significance." *Palaios* 27, no. 9: 594–606.

Moysiuk, J., M. R. Smith, and J.-B. Caron. 2017.

"Hyoliths Are Palaeozoic Lophophorates." *Nature* 541:394–397.

Robison, R. A., and L. E. Babcock. 2011. "Systematics, Paleobiology, and Taphonomy of Some Exceptionally Preserved Trilobites from Cambrian Lagerstätten of Utah." *Paleontological Contributions* 5:1–47.

Robison, R. A., L. E. Babcock, and V. G. Gunther. 2015. *Exceptional Cambrian Fossils from Utah: A Window into the Age of Trilobites*. Utah Geological Survey Miscellaneous Publication 15-1.

Skovsted, C. B., and L. E. Holmer. 2006. "The Lower Cambrian Brachiopod *Kyrshabaktella* and Associated Shelly Fossils from the Harkless Formation, Southern Nevada." *GFF (Journal of the Geological Society of Sweden)* 128, no. 4:327–337.

Sappenfield, A. D. 2015. "Precambrian-Cambrian Sedimentology, Stratigraphy, and Paleontology in the Great Basin (Western United States)." PhD diss., University of California, Riverside. https://escholarship.org/uc/item/4ro2d6xr.

Webster, M., R. R. Gaines, N. C. Hughes. 2008. "Microstratigraphy, Trilobite Biostratinomy, and Depositional Environment of the "Lower Cambrian" Ruin Wash Lagerstätte, Pioche Formation, Nevada." *Palaeogeography, Palaeoclimatology, Palaeoecology* 264:100–122.

Zhang, F., S. Xiao, B. Kendall, S. J. Romaniello, H. Cui, M. Meyer, G. J. Gilleaudeau, A. J. Kaufman, and A. D. Anbar. 2018. "Extensive Marine Anoxia during the Terminal Ediacaran Period." *Science Advances* 4, no. 6:1–11.

Chapter 4. The Ordovician Overhaul

Berry, W. B. N., R. L. Ripperdan, and S. C. Finney. 2002. "Late Ordovician Extinction: A Laurentian View." In *Catastrophic Events and Mass Extinctions: Impacts and Beyond*, edited by C. Koeberl and K. G. MacLeod, 463–471. Geological Society of America Special Paper 356.

Bond, D. P. G., and S. E. Grasby. 2020. "Late Ordovician Mass Extinction Caused by Volcanism, Warming, and Anoxia, Not Cooling and Glaciation." *Geology* 48:777–781.

Boyer, D., and M. L. Droser. 2003. "Shell Beds of the Kanosh and Lehman Formations of Western Utah: Paleoecological and Paleoenvironmental Interpretations." *Brigham Young University Geology Studies* 47:1–15.

Carrera, M. G., and J. K. Rigby. 1999. "Biogeography of Ordovician Sponges." *Journal of Paleontology* 73, no. 1:26–37.

Church, S. B. 1991. "A New Lower Ordovician Species of *Calathium*, and Skeletal Structure of Western Utah Calathids." *Journal of Paleontology* 65, no. 4:602–610.

———. 2017. "Efficient Ornamentation in Ordovician Anthaspidellid Sponges." *Paleontological Contributions* 18:1–8.

Cooper, J. D., and M. K. Keller. 2001. "Paleokarst in the Ordovician of the Southern Great Basin, USA: Implications for Sea-Level History." *Sedimentology* 48:855–873.

Dahl, R. M. 2012. "A Paleoenvironmental Analysis of Gastropods from the Middle Ordovician, Ibex Region, Utah." Master's thesis, University of California, Riverside. https://escholarship.org/uc/item/2227n908Sappenfield.

Delabroye, A., and M. Vecoli. 2010. "The End-Ordovician Glaciation and the Hirnantian Stage: A Global Review and Questions about Late Ordovician Event Stratigraphy." *Earth Science Reviews* 98, nos. 3–4:269–282.

Dilly, N. 2009. "*Cephalodiscus graptolitoides* sp. nov. a Probable Extant Graptolite." *Journal of Zoology* 229, no. 1:69–78.

Droser, M., and S. Finnegan. 2003. "The Ordovician Radiation: A Follow-Up to the Cambrian Explosion." *Integrative and Comparative Biology* 43:178–184.

Druschke, A., G. Jiang, T. B. Anderson, and A. D. Hanson. 2009. "Stromatolites in the Late Ordovician Eureka Quartzite: Implications for Microbial Growth and Preservation in Siliciclastic Settings." *Sedimentology* 56:1275–1291.

Edwards, C. T., M. R. Saltzman, D. L. Royer, and D. A. Fike. 2017. "Oxygenation as a Driver of the Great Ordovician Biodiversification Event." *Nature Geoscience* 10:925–929.

Finnegan, S., and M. Droser. 2005. "Relative and Absolute Abundance of Trilobites and Rhynchonelliform Brachiopods across the Lower/Middle Ordovician Boundary, Eastern Basin and Range." *Paleobiology* 31, no. 3:480–502.

———. 2008. "Reworking Diversity: Effects of Storm Deposition on Evenness and Sampled

Richness, Ordovician of the Basin and Range, Utah and Nevada, USA." *Palaios* 23, no. 2: 87–96.

Finnegan, S., S. Peters, and W. W. Fischer. 2011. "Late Ordovician-Early Silurian Selective Extinction Patterns in Laurentia and Their Relationship to Climate Change." In *Ordovician of the World*, edited by J. C. Gutiérrez-Marco, I. Rábano, and D. García-Bellido, 3–9. Cuadernos del Museo Geominero 14. Madrid: Instituto Geológico y Minero de España.

Finney, S. C., W. B. N. Berry, J. D. Cooper, R. L. Ripperdan, W. C. Sewwt, S. R. Jacobson, A. Soufiane, A. Achab, and P. J. Noble. 1999. "Late Ordovician Mass Extinction: A New Perspective from Stratigraphic Sections in Central Nevada." *Geology* 27:215–218.

Finney, S. C., and R. L. Ethington. 2000. "Global Ordovician Series Boundaries and Global Event Biohorizons, Monitor Range and Roberts Mountains, Nevada." In *Geological Society of America Field Guide 2: Great Basin and Sierra Nevada*, edited by D. R. Lageson, S. G. Peters, and M. M. Lahren, 301–318. Boulder, CO: Geological Society of America.

Fortey, R. A., and M. L. Droser. 1999. "Trilobites from the Base of the Type Whiterockian (Middle Ordovician) in Nevada." *Journal of Paleontology* 73, no. 2:182–201.

French, B. M., R. M. McKay, H. P. Liu, D. E. G. Briggs, and B. J. Witzke. 2018. "The Decorah Structure, Northeastern Iowa: Geology and Evidence for Formation by Meteorite Impact." *Geological Society of America Bulletin* 139, no. 11–12:2062–2086.

Henry, S. E. 2014. "A New Species of an Enigmatic Fossil Taxon: *Ischadites* n. sp., a Middle Ordovician Receptaculitid from the Great Basin, Western USA." Master's thesis, University of California Riverside. ProQuest (Henry_ucr_0032N_11862), https//escholarship.org/uc/item/25q6j3nz.

Johns, R. A. 1994. "Ordovician Lithistid Sponges of the Great Basin." *Nevada Bureau of Mines and Geology Open-File Report* 94-1.

Keller, M., and J. Cooper. 1995. "Paleokarst in the Lower Middle Ordovician of Southeastern California and Adjacent Nevada and Its Bearing on the Sauk-Tippecanoe Boundary

Problem." In *Ordovician Odyssey: Short Papers for the 7th International Symposium on the Ordovician System*, edited by J. D. Cooper, M. L. Droser, and S. L. Finney, 323–327. Society for Sedimentary Geology (SEPM), Pacific Section 77.

Keller, M., and O. Lehnert. 2010. "Ordovician Paleokarst and Quartz Sand: Evidence of Volcanically-Triggered Extreme Climates." *Palaeogeography, Palaeoclimatology, Palaeoecology* 296:297–309.

Kervin, R. J., and A. D. Woods. 2012. "Origin and Evolution of Palaeokarst within the Lower Ordovician (Ibexian) Goodwin Formation (Pogonip Group)." *Journal of Palaeogeography* 1, no. 1:57–69.

Ketner, K. B. 1968. "Origin of Ordovician Quartzite in the Cordilleran Miogeosyncline." U.S. Geological Survey Professional Paper 600-B, 169–177.

———. "Deposition and Deformation of Lower Paleozoic Western Facies Rocks, Northern Nevada." In *Paleozoic Paleogeography of the Western United States*, edited by J. H. Stewart, C. H. Stevens, and A. E. Fritsche, 251–258. Los Angeles: Pacific Section of the Society of Economic Paleontologists and Mineralogists.

Kim, J. C., and Y. I. Lee. 1995. "Flat-pebble Conglomerate: A Characteristic Lithology of Upper Cambrian and Lower Ordovician Shallow-Water Carbonate Sequences." In *Ordovician Odyssey: Short Papers for the 7th International Symposium on the Ordovician System*, edited by J. D. Cooper, M. L. Droser, and S. L. Finney, 371–374. Society for Sedimentary Geology (SEPM), Pacific Section 77.

Lee, J.-H., B. F. Dattilo, J. F. Miller, S. Mrozek, and R. Riding. 2017. "Lithistid Sponge-Microbial Reefs, Nevada, USA. Filling the Late Cambrian 'Reef Gap.'" *Palaegeography, Palaeoclimatology, Palaeoecology* 520:251–262.

Li, X., and M. L. Droser. 1999. "Lower and Middle Ordovician Shell Beds from the Basin and Range Province of the Western United States (California, Nevada, and Utah)." *Palaios* 14:215–233.

Marenco, K. N. 2017. "Reef-Building at the Dawn of the GOBE: The Rise of Metazoan Framework Constituents in Lower Ordovician

Reefs, Western Utah, USA." *Geological Society of America, Abstracts with Programs* 49, no. 6, paper 132-7.

Meinhold, G., A. Arslan, O. Lehnert, and G. M. Stampfli. 2011. "Global Mass Wasting during the Middle Ordovician: Meteoritic Trigger or Plate-Tectonic Environment?" *Gondwana Research* 19, no. 2:535–541.

Ormö, J., E. Sturkell, C. Awlmark, and J. Melosh. 2014. "First Known Terrestrial Impact of a Binary Asteroid from a Main Belt Breakup Event." *Nature Scientific Reports* 4, article 6724.

Pohl, A., and J. Austermann. 2018. "A Sea-Level Fingerprint of the Late Ordovician Ice-Sheet Collapse." *Geology* 46, no. 7:595–598.

Pruss, S. B., S. Finnegan, W. W. Fischer, and A. H. Knoll. 2010. "Carbonates in Skeleton-Poor Seas: New Insights from Cambrian and Ordovician Strata of Laurentia." *Palaios* 25, no. 2:73–84.

Regenfuss, S. M., J. M. Strickland, and J. D. Cooper. 1999. "The Mystery of the Eureka Breccia." *California Geology* 52, no. 3:4–15.

Rigby, J. K., and P. Jamison. 1994. "Lithistid Sponges from the Late Ordovician Fish Haven Dolomite, Bear River Range, Cache County, Utah." *Journal of Paleontology* 68, no. 4: 722–726.

Rohr, D. 1994. "Ordovician (Whiterockian) Gastropods of Nevada: Bellerophontoidea, Macluritoidea, and Euomphaloidea." *Journal of Paleontology* 68, no. 3:473–486.

Ross, R. J., Jr. 1970. "Ordovician Brachiopods, Trilobites, and Stratigraphy in Eastern and Central Nevada." U.S. Geological Survey Professional Paper 639.

Ross, R. J., Jr., and C. R. Longwell. 1964. "Middle and Lower Ordovician Formations in Southernmost Nevada and Adjacent California." *U.S. Geological Survey Bulletin* 1180-C.

Ross, R. J., Jr., L. F. Hintze, R. L. Ethington, J. F. Miller, M. E. Taylor, J. E. Repetski, J. E. Sprinkle, and T. E. Guensburg. 1993. "The Ibexian Series (Lower Ordovician), a Replacement for the 'Canadian Series' in North American Chronostratigraphy." U.S. Geological Survey Open-File Report 93-598.

Saltzman, M. R., and S. A. Young. 2005. "Long-Lived Glaciation in the Late Ordovician?

Isotopic and Sequence-Stratigraphic Evidence from Western Laurentia." *Geology* 33, no. 2: 109–112.

Schmitz, B., D. Harper, B. Peuker-Ehrenbrink, S. Stouge, C. Alwmark, A. Cronholm, S. Bergström, M. Tassinari, and Wang Xiaofeng. 2007. "Asteroid Breakup Linked to the Great Ordovician Biodiversification Event." *Nature Geoscience* 1:49–53.

Scotese, C. R. 2014. *Atlas of Silurian and Middle–Late Ordovician Paleogeographic Maps (Mollweide Projection)*, maps 73–80. Vol. 5, *The Early Paleozoic, PALEOMAP Atlas for ArcGIS*. Evanston, IL: PALEOMAP Project.

Scotese, C. R., and N. Wright. 2018. PALEOMAP Paleodigital Elevation Models (PaleoDEMS) for the Phanerozoic PALEOMAP Project, https://www.earthbyte.org/paleodem -resourcescotese-and-wright-2018.

Servais, T. M., and D. A. T. Harper. 2018. "The Great Ordovician Biodiversification Event (GOBE): Definition, Concept, and Duration." *Lethaia* 51:151–164.

Sloss, L. L. 1963. "Sequences in the Cratonic Interior of North America." *Geological Society of America Bulletin* 74:93–113.

Young, S. A., M. R. Saltzman, W. I. Ausich, A. Desrochers, and D. Kaljo. 2010. "Did Changes in Atmospheric CO_2 Coincide with Latest Ordovician Glacial-Interglacial Cycles?" *Palaeogeography, Palaeoclimatology, Palaeoecology* 296:376–388.

Zimmerman, M. K., and J. D. Cooper. 1999. "Sequence Stratigraphy of the Middle Ordovician Eureka Quartzite, Southeastern California and Southern Nevada, USA." In *Quo Vadis Ordovician? Short Papers for the 8th International Symposium on the Ordovician System*, edited by P. Kraft and O. Fatka. *Acta Universitatis Carolinae (Geologica)* 43, nos. 1–2:147–150.

Chapter 5. The Dolomite Interval: A Carbonate Conundrum

Algeo, T. J., R. A. Berner, J. P. Maynard, and S. E. Scheckler. 1995. "Late Devonian Oceanic Anoxic Events and Biotic Crises: 'Rooted' in the Evolution of Vascular Land Plants?" *GSA Today* 5:64–66.

Barash, M. S. 2016. "Causes of the Great Mass Extinction of Marine Organisms in the Late Devonian." *Oceanology* 56, no. 6:863–875.

Bond, D., and P. B. Wignall. 2005. "Evidence for Late Devonian (Kellwasser) Anoxic Events in the Great Basin, Western United States." In *Developments in Palaeontology and Stratigraphy*. Vol. 20, Understanding Late Devonian and Permian-Triassic Biotic and Climatic Events: Towards an Integrated Approach, edited by D. J. Over, J. R. Morrow, and P. B. Wignall, 225–262. Elsevier Science.

Bontognali, T. R. R. 2019. "Anoxygenic Phototrophs and the Forgotten Art of Making Dolomite." *Geology* 47:591–592.

Burrow, C. J. 2003. "Poracanthodid Acanthodian from the Upper Silurian (Pridoli) of Nevada." *Journal of Vertebrate Paleontology* 23, no. 3: 489–493.

———. 2007. "Early Devonian (Emsian) Ancanthodian Fish Faunas of the Western USA." *Journal of Paleontology* 81, no. 5: 824–840.

Burrow, C. J., and M. A. Murphy. 2016. "Early Devonian (Pragian) Vertebrates from the Northern Roberts Mountains, Nevada." *Journal of Paleontology* 90, no. 4:734–740.

Dunham, J. B., and E. R. Olson. 1980. "Shallow Subsurface Dolomitization of Subtidally Deposited Carbonate Sediments in the Hanson Creek Formation (Ordovician-Silurian) of Central Nevada." In *Concepts and Models of Dolomitization*, edited by D. H. Zenger, J. B. Dunham, and R. L. Ethington. SEPM (Society for Sedimentary Geology) Special Publication 28:139–162.

Elliott, D. K. 1994. "New Pteraspidid (Agnatha: Heterostraci) from the Lower Devonian Water Canyon Formation of Utah." *Journal of Paleontology* 68, no. 1:176–179.

Elliott, D. K., and A. R. M. Blieck. 2010. "A New Ctenaspid (Agnatha, Heterostraci) from the Early Devonian of Nevada, with Comments on Taxonomy, Paleobiology, and Paleobiogeography." In *Morphology, Phylogeny, and Paleobiogeography of Fossil Fishes*, edited by D. K. Elliott, J. G. Maisey, X. Yu, and D. Miao, 25–38. Munich: Verlag Dr. Friedrich Pfeil.

Elliott, D. K., and R. R. Ilyes. 1996. "New Early Devonian Pteraspidids (Agnatha, Heterostraci) from Death Valley National Monument, Southeastern California." *Journal of Paleontology* 70, no. 1:152–161.

Elliot, D. K., and H. G. Johnson. 1997. "Use of Vertebrates to Solve Biostratigraphic Problems: Examples from the Lower and Middle Devonian of Western North America." In *Paleozoic Sequence Stratigraphy, Biostratigraphy, and Biogeography: Studies in Honor of J. Granville ("Jess") Johnson*, edited by G. Klapper, M. A. Murphy, and J. A. Talent, 179–188. Geological Society of America Special Paper 321.

Elliott, D. K., and M. A. Petriello. 2011. "New Poraspids (Agnatha, Heterostraci) from the Early Devonian of the Western United States." *Journal of Vertebrate Paleontology* 31:518–531.

Elliott, D. K., R. C. Reed, and H. G. Johnson. 1999. "The Devonian Vertebrates of Utah." In *Vertebrate Paleontology in Utah*, edited by D. Gillette. Utah Geological Survey, Miscellaneous Publication 99-1:1–12.

Frederick, A., and D. M. Rohr. 2015. "First Reported Occurrence of Late Ordovician through Early Silurian Gastropod Genera: *Phragmolites, Oriostoma, Pachystrophia, Euomphalopterus, Onychochilus,* and *Platyostoma* from the Eastern Great Basin of North America." In *Fossil Record 4*, edited by R. M. Sullivan and S. G. Lucas, 51–57. *New Mexico Museum of Natural History and Science Bulletin 67*.

Harris, M. T., and P. M. Sheehan. 1997. "Carbonate Sequences and Fossil Communities from Then Upper Ordovician-Lower Silurian of the Eastern Great Basin." In *Proterozoic to Recent Stratigraphy, Tectonics, and Volcanology, Utah, Nevada, Southern Idaho, and Central Mexico*, edited by P. K. Link and B. J. Kowallis. *Brigham Young University Geology Studies* 42:105–128.

Johnson, H. G., D. K. Elliott, and J. H. Wittke. 2000. "A New Actinolepid Arthrodire (Class Placodermi) from the Lower Devonian Sevy Dolomite, East-Central Nevada." *Zoological Journal of the Linnean Society* 129, no. 2: 241–266.

Johnson, J. G., G. Klapper, and J. G. Johnson. 1990. "Lower and Middle Devonian Brachiopod-Dominated Communities of Nevada, and Their Position in a Biofacies-Province-Realm Model." *Journal of Paleontology* 64, no. 6:902–941.

Johnson, J. G., and M. A. Murphy. 1984. "Time-Rock Model for Siluro-Devonian Continental Shelf, Western United States." *Geological Society of America Bulletin* 95:1349–1359.

Krause, S., V. Liebtrau, S. Gorb, M. Sanchez-Roman, J. A. McKenzie, and T. Treude. 2012. "Microbial Nucleation of Mg-Rich Dolomite in Exopolymeric Substances under Anoxic Modern Seawater Salinity: New Insights into an Old Enigma." *Geology* 40, no. 7:587–590.

Land, L. S. 1985. "The Origin of Massive Dolomite." *Journal of Geological Education* 33, no. 2:112–125.

Matti, J. C., M. A. Murphy, and S. C. Finney. 1975. "Silurian and Lower Devonian Basin and Slope Limestones, Copenhagen Canyon, Nevada." Geological Society of America Special Paper 159.

McGhee, G. 1996. *The Late Devonian Mass Extinction: The Frasnian/Famennian Crisis.* New York: Columbia University Press.

Merriam, C. W. 1973. "Paleontology and Stratigraphy of the Rabbit Hill Limestone and Lone Mountain Dolomite of Central Nevada." U.S. Geological Survey Professional Paper 808.

———. 1963. "Paleozoic Rocks of Antelope Valley, Eureka and Nye Counties, Nevada." U.S. Geological Survey Professional Paper 423.

———. 1973. "Silurian Rugose Corals of the Central and Southwest Great Basin." U.S. Geological Survey Professional Paper 777.

Merriam, C. W., and E. H. McKee. 1976. "The Roberts Mountains Formation, a Regional Stratigraphic Study with Emphasis on Rugose Coral Distribution." U.S. Geological Survey Professional Paper 973.

Miller, R. H. 1976. "Revision of Upper Ordovician, Silurian, and Lower Devonian Stratigraphy, Southwestern Great Basin." *Geological Society of America Bulletin* 87, no. 7:961–968.

Munnecke, A., M. Calner, D. A. T. Harper, and T. Servais. 2010. "Ordovician and Silurian

Sea-Water Chemistry, Sea Level, and Climate: A Synopsis." *Palaeogeography, Palaeoclimatology, Palaeoecology* 296:389–413.

Osmund, J. C. 1962. "Stratigraphy of the Devonian Sevy Dolomite in Utah and Nevada." *American Association of Petroleum Geologists Bulletin* 46, no. 11:2033–2056.

Ryb, U., and J. M. Eiler. 2018. "Oxygen Isotope Composition of the Phanerozoic Ocean and a Possible Solution to the Dolomite Problem." *Proceedings of the National Academy of Sciences* 115, no. 26:6602–6607.

Sallan, L. C., and M. I. Coates. 2010. "End-Devonian Extinction and a Bottleneck in the Early Evolution of Modern Jawed Vertebrates." *Proceedings of the National Academy of Sciences* 107, no. 22:10131–10135.

Saltzman, M. R. 2002. "Carbon Isotope ($\delta^{13}C$) Stratigraphy across the Silurian-Devonian Transition in North America: Evidence for a Perturbation of the Global Carbon Cycle." *Palaeogeography, Palaeoclimatology, Palaeoecology* 187:83–100.

Sandberg, C. A., J. R. Morrow, and W. Ziegler. 2002. "Late Devonian Sea-Level Changes, Catastrophic Events, and Mass Extinctions." In *Catastrophic Events and Mass Extinctions: Impacts and Beyond*, edited by C. Koeberl and K. G. MacLeod, 473–487. Geological Society of America Special Paper 356.

Schultze, H.-P. 2010. "The Late Middle Devonian Fauna of Red Hill I, Nevada, and Its Paleobiogeographic Implications." *Fossil Record* 13, no. 2:285–295.

Sheehan, M. 1979. "Silurian Continental Margin in Northern Nevada and Northwest Utah." *University of Wyoming Contributions to Geology* 17, no. 1:25–35.

Stigall, A. L. 2012. "Speciation Collapse and Invasive Species Dynamics during the Late Devonian 'Mass Extinction.'" *GSA Today* 22, no. 1:4–9.

Swartz, B. 2012. "A Marine Stem-Tetrapod from the Devonian of Western North America." *PLoS ONE* 7(3):e33683. https://doi.org/10.1371/journal.pone.0033683.

Rigby, J. K. 1967. "Sponges from the Silurian Laketown Dolomite, Confusion Range, West-

ern Utah." *Brigham Young University Geology Studies* 14:241–258.

Turner, S. T., and M. A. Murphy. 1988. "Early Devonian Microfossils from the Simpson Park Range, Eureka County, Nevada." *Journal of Paleontology* 62, no. 6:959–964.

Vodrážková, S., G. Klapper, and M. A. Murphy. 2011. "Early Middle Devonian Conodont Faunas (Eifelian, Costatus–Kockelianus Zones) from the Roberts Mountains and Adjacent Areas in Central Nevada." *Czech Geological Survey Bulletin of Geosciences* 86, no. 4:737–764.

Winterer, E. L., and M. A. Murphy. 1960. "Silurian Reef Complex and Associated Facies, Central Nevada." *Journal of Geology* 68, no. 2:117–139.

Chapter 6. The Devonian-Mississippian Interval: Tectonic Tumult and Cosmic Calamities on the Sagebrush Seafloor.

Burchfiel, B. C., and L. H. Royden. 1991. "Antler Orogeny: A Mediterranean-Type Orogeny." *Geology* 19:66–69.

Carroll, R. L., P. Bybee, and W. D. Tidwell. 1991. "The Oldest Microsaur (Amphibia)." *Journal of Paleontology* 65, no. 2:314–322.

Claeys, P., J.-G. Casier, and S. V. Margolis. 1992. "Microtektites and Mass Extinctions: Evidence for a Late Devonian Asteroid Impact." *Science* 257:1102–1104.

Cole, D., P. M. Myrow, D. A. Fike, A. Hakim, and G. E. Gehrels. 2015. "Uppermost Devonian (Famennian) to Lower Mississippian Events of the Western U.S.: Stratigraphy, Chemostratigraphy, and Detrital Zircon Geochronology." *Palaeogeography, Palaeoclimatology, Palaeoecology* 427:1–19.

Dickinson, W. R. 1977. "Paleozoic Plate Tectonics and the Evolution of the Cordilleran Continental Margin." In *Paleozoic Paleogeography of the Western United States: Pacific Coast Paleogeography Symposium 1*, edited by J. H. Stewart, C. H. Stevens, and A. E. Fritsche, 137–155. Los Angeles: Pacific Section, Society of Economic Paleontologists and Mineralogists.

Evans, J. G., and T. G. Theodore. 1978. "Deformation of the Roberts Mountains Allochthon in North-Central Nevada." U.S. Geological Survey Professional Paper 1060.

Gehrels, G. E., and W. R. Dickinson. 2000. "Detrital Zircon Geochronology of the Antler Overlap and Foreland Basin Assemblage, Nevada." In *Paleozoic and Triassic Paleogeography and Tectonics of Western Nevada and Northern California*, edited by M. J. Soreghan and G. E. Gehrels, 57–64. Geological Society of America Special Paper 347.

Johnson, J. G., and A. Pendergast. 1981. "Timing and Mode of Emplacement of the Roberts Mountains Allochthon, Antler Orogeny." *Geological Society of America Bulletin* 92, no. 9:648–658.

Jewell, W., N. J. Silberling, and K. M. Nichols. 2000. "Geochemistry of the Mississippian Delle Phosphatic Event, Eastern Great Basin, U.S.A." *Journal of Sedimentary Research* 70, no. 5:1222–1233.

Ketner, K. B. 2012. "An Alternative Hypothesis for the Mid-Paleozoic Antler Orogeny in Nevada." U.S. Geological Survey Professional Paper 1790.

———. 2013. "Stratigraphy of Lower to Middle Paleozoic Rocks of Northern Nevada and the Antler Orogeny." U.S. Geological Survey Professional Paper 1799.

Ketner, K. B., and J. F. Smith Jr. 1982. "Mid-Paleozoic Age of the Roberts Thrust Unsettled by New Data from Northern Nevada." *Geology* 10:298–303.

Mattinson, C. G., and B. H. Tiffney. 2001. "Terrestrial Plant Fossils of the Mississippian Diamond Peak Formation, White Pine Range, Eastern Nevada." *Paleobios* 21, no. 3:1–11.

Mickle, K. E. 2011. "The Early Actinopterygian Fauna of the Manning Canyon Shale (Upper Mississippian, Lower Pennsylvanian) of Utah, U.S.A." *Journal of Vertebrate Paleontology* 31, no. 5:962–980.

Miller, E. L., M. M. Miller, C. H. Stevens, J. E. Wright, and R. Madrid. 1992. "Late Paleozoic Paleogeographic and Tectonic Evolution of the Western U.S. Cordillera." In *The Geology of North America*. Vol. G-3, *The Cordilleran Orogen: Conterminous U.S.*, edited by B. C. Burchfiel, P. W. Lipman, and

M. L. Zoback, 57–106. Boulder, CO: Geological Society of America.

Miller, W. E. 1981. "Cladodont Shark Teeth from Utah." *Journal of Paleontology* 55:894–895.

Morrow, J. R., and C. A. Sandberg. 2008. "Evolution of Devonian Carbonate Shelf Margin, Nevada." *Geosphere* 4, no. 2:445–458.

Morrow, J. R., C. A. Sandberg, and A. G. Harris. 2005. "Late Devonian Alamo Impact, Southern Nevada, USA: Evidence of Size, Marine Site, and Widespread Effects." In *Large Meteorite Impacts III*, edited by T. Kenkmann, F. Hörz, and A. Deutsch, 259–280. Geological Society of America Special Paper 384.

Morrow, J. R., C. A. Sandberg, K. Malkowski, and M. M. Joachimski. 2009. "Carbon Isotope Chemostratigraphy and Precise Dating of Middle Frasnian (Lower Upper Devonian) Alamo Breccia, USA." *Palaeogeography, Palaeoclimatology, Palaeoecology* 282:105–118.

Nilsen, T. H., and J. H. Stewart, conveners. 1980. "The Antler Orogeny-Mid-Paleozoic Tectonism in Western North America." *Geology* 8: 298–302.

Pinto, J. A., and J. E. Warme. 2008. "Alamo Event, Nevada: Crater Stratigraphy and Impact Breccia Realms." In *The Sedimentary Record of Meteorite Impacts*, edited by K. R. Evans, J. W. Horton Jr., D. T. King Jr., and J. R. Morrow. Geological Society of America Special Paper 437:99–137.

———. 2016. "Seismite in the Devonian Sultan Formation, Frenchman Mountain, Nevada: Evidence of Far-Field Effect of the Alamo Event." *Mountain Geologist* 53, no. 2:93–14.

Poole, F. G., and C. A. Sandberg. 2015. "Alamo Impact Olistoliths in Antler Orogenic Foreland, Warm Springs-Milk Spring area, Hot Creek Range, Central Nevada." In *Unusual Central Nevada Terranes Produced by Late Devonian Antler Orogeny and Alamo Impact*, by F. G. Poole and C. A. Sandberg, 39–104. Geological Society of America Special Paper 517.

———. 2015. "Olistrostome Shed Eastward from the Antler Orogenic Forebulge, Bisoni-McKay Area, Fish Creek Range, Central Nevada." In *Unusual Central Nevada Terranes Produced by Late Devonian Antler Orogeny and Alamo Impact*, by F. G. Poole and C. A. Sandberg, 1–38. Geological Society of America Special Paper 517.

Rendall, B. E., and L. Tapanila. 2020. "Impact Resilience: Ecological Recovery of a Carbonate Factory in the Wake of the Late Devonian Impact Event." *Palaios* 53, no. 1:12–21.

Retzler, A. J., L. Tapanila, J. R. Steenberg, C. J. Johnson, and R. A. Myers. 2015. "Post-Impact Depositional Environments as a Proxy for Crater Morphology, Late Devonian Alamo Impact, Nevada." *Geosphere* 11, no. 1:123–143.

Roberts, R. J. 1964. "Stratigraphy and Structure of the Antler Peak Quadrangle, Humboldt and Lander Counties, Nevada." U.S. Geological Survey Professional Paper 459-A:A1–A93.

Saltzman, M. R. 2003. "Organic Carbon Burial and Phosphogenesis in the Antler Foreland Basin: An Out-of-Phase Relationship during the Lower Mississippian." *Journal of Sedimentary Research* 73, no. 6:844–855.

Saltzman, M. R., L. A. Gonzalez, and K. C. Lohmann. 2000. "Earliest Carboniferous Cooling Step Triggered by the Antler Orogeny?" *Geology* 28, no. 4:347–350.

Sandberg, C. A., J. R. Morrow, and J. E. Warme. 1997. "Late Devonian Alamo Impact Event, Global Kellwasser Events, and Major Eustatic Events, Eastern Great Basin, Nevada and Utah." *Brigham Young University Geology Studies* 42, no. 1:129–160.

Sandberg, C. A., F. G. Poole, and J. R. Morrow. 2001. "Construction and Destruction of Crinoidal Mudmounds on Mississippian Antler Forebulge, East of Eureka, Nevada." In *Structure and Stratigraphy of the Eureka, Nevada Area*, edited by M. S. Miller, and J. P. Walker, 23–50. Nevada Petroleum and Geothermal Society 2001 Fieldtrip Guidebook NPS16y.

Sando, W. J. 1984. "Significance of Epibionts on Horn Corals from the Chainman Shale (Upper Mississippian) of Utah." *Journal of Paleontology* 58, no. 1:185–196.

Speed, R. C., and N. H. Sleep. 1982. "Antler Orogeny and Foreland Basin: A Model." *Geological Society of America Bulletin* 93:815–828.

Stahl, S. D. 1989. "Recognition of Jurassic

Transport of Rocks of the Roberts Mountains Allochthon: Evidence from the Sonoma Range, North-Central Nevada." *Geology* 17: 645–648.

Stevens, C. H., and T. Pelley. 2006. "Development and Dismemberment of a Middle Devonian Continental-Margin Submarine Fan System in East-Central California." *Geological Society of America Bulletin* 118, nos. 1–2:159–170.

Tapanila, L., J. R. Steenberg, C. J. Johnson, and R. A. Myers. 2014. "The Seafloor after a Bolide Impact: Sedimentary and Biotic Signatures across the Late Devonian Carbonate Platform following the Alamo Impact Event, Nevada, USA." *Facies* 60:615–629.

Thomson, K. S., N. S. Shubin, and F. G. Poole. 1998. "A Problematic Early Tetrapod from the Mississippian of Nevada." *Journal of Vertebrate Paleontology* 18, no. 2:315–320.

Tidwell, W. D., D. A. Medlyn, and D. A. Simper. 1974. "Flora of the Manning Canyon Shale, Part II: Lepidodendrales." *Brigham Young University Geology Studies* 21:119–146.

Tidwell, W. D., J. R. Jennings, and B. Call. 1988. "Flora of the Manning Canyon Shale, Part III: Sphenophyta." *Brigham Young University Geology Studies* 35:15–32.

Titus, A. L. 2000. "Late Mississippian (Arnsbergian Stage-E_2 Chronozone) Ammonoid Paleontology and Biostratigraphy of the Antler Foreland Basin, California, Nevada, Utah." *Utah Geological Survey Bulletin* 131.

Trexler, J. H., Jr., P. H. Cashman, W. S. Snyder, and I. Davydov. 2004. "The Western Margin of North America after the Antler Orogeny: Mississippian through Late Permian History in the Basin and Range, Nevada." In *Geological Field Trips in Southern Idaho, Eastern Oregon, and Northern Nevada*, edited by K. M. Haller and S. H. Wood, 20–37. U.S. Geological Survey Open-File Report 2004-1222.

Vodrážková, S., G. Klapper, and M. A. Murphy. 2011. "Early Middle Devonian Conodont Faunas (Eifelian, Costatus–Kockelianus Zones) from the Roberts Mountains and adjacent areas in Central Nevada." *Czech Geological Survey Bulletin of Geosciences* 86, no. 4:737–764.

Warme, J. E., and H. C. Kuehner. 1998. "Anatomy of an Anomaly: The Devonian Catastrophic Alamo Impact Breccia of Southern Nevada." *International Geology Review* 40:189–216.

Warme, J. E., and C. A. Sandberg. 1996. "Alamo Megabreccia: Record of a Late Devonian Impact in Southern Nevada." *GSA Today* 6, no. 1:1–7.

Chapter 7. The Antler Overlap Sequence

Benton, M. J. 2018. "Hyperthermal-Driven Mass Extinctions: Killing Models during the Permian-Triassic Mass Extinction." *Philosophical Transactions of the Royal Society A* 376:20170076.

Berger I., D. A. Singer, and T. G. Theodore. 2001. "Sedimentology of the Pennsylvanian and Permian Strathearn Formation, Northern Carlin Trend, Nevada." U.S. Geological Survey Open-File Report 2011-401.

Bishop, J. W., I. P. Montañez, E. L. Gulbranson, and L. Brenckle. 2009. "The Onset of Mid-Carboniferous Glacio-Eustasy: Sedimentology and Diagenetic Constraints, Arrow Canyon, Nevada." *Palaeogeography, Palaeoclimatology, Palaeoecology*, 276, nos. 1–4: 217–243.

Bishop, J. W., I. P. Montañez, and D. A. Osleger. 2010. "Dynamic Carboniferous Climate Change, Arrow Canyon, Nevada." *Geosphere* 6, no. 1:1–34.

Brand, U., and P. Brenckle. 2001. "Chemostratigraphy of the Mid-Carboniferous Boundary Global Stratotype Section and Point (GSSP), Bird Spring Formation, Arrow Canyon, Nevada, USA." *Palaeogeography, Palaeoclimatology, Palaeoecology* 165, nos. 3–4: 321–347.

Brueckner, H. K., and W. S. Snyder. 1985. "Structure of the Havallah Sequence, Golconda Allochthon, Nevada: Evidence for Prolonged Evolution in an Accretionary Prism." *Geological Society of America Bulletin* 96:1113–1130.

Burger, B., M. V. Estrada, and M. S. Gustin. 2019. "What Caused Earth's Largest Mass Extinction Event? New Evidence from the Permian-Triassic Boundary in Northeastern Utah." *Global and Planetary Change* 177:81–100.

Burgess, S. D., S. Bowring, and S.-Z. Shen. 2014. "High-Precision Timeline for Earth's Most Severe Extinction." *Proceedings of the National Academy of Sciences* 111:3316–3321.

Cashman, P., J. Trexler, W. Snyder, V. Davydov, and W. Taylor. 2008. "Late Paleozoic Deformation in Central and Southern Nevada." In *Geological Society of America Field Guide 11: Field Guide to Plutons, Volcanoes, Faults, Reefs, Dinosaurs, and Possible Glaciation in Selected Areas of Arizona, California, and Nevada,* edited by E. M. Duebendorfer and E. I. Smith, 21–42. Boulder, CO: Geological Society of America.

Carroll. A. R., N. P. Stephens, M. S. Hendrix, and C. R. Glenn. 1998. "Eolian-Derived Siltstone in the Upper Permian Phosphoria Formation: Implications for Marine Upwelling." *Geology* 26, no. 11:1023–1026.

Cassity, E., and R. L. Langenheim Jr. 1966. "Pennsylvanian and Permian Fusulinids of the Bird Spring Group from Arrow Canyon, Clark County, Nevada." *Journal of Paleontology* 40:931–968.

Cecil, C. B. 2015. "Paleoclimate and the Origin of Paleozoic Chert in the Rock Record." *The Sedimentary Record* 13, no. 3:1–10.

Crafford, A. E. J. 2008. "Paleozoic Tectonic Domains of Nevada: An Interpretive Discussion to Accompany the Geologic Map of Nevada." *Geosphere* 4, no. 1:260–2

Dickinson, W. R. 2013. "Phanerozoic Palinspastic Reconstructions of Great Basin Geotectonics (Nevada-Utah, USA)." *Geosphere* 9, no. 5: 1384–1396.

Erskine, M. C. 1997. "The Oquirrh Basin Revisited." *American Association of Petroleum Geologists Bulletin* 81, no. 4:624–636.

Fielding, C. R., T. D. Frank, and J. L. Isbell. 2008. "The Late Paleozoic Ice Age-A Review of Current Understanding and Synthesis of Global Climate Patterns." In *Resolving the Late Paleozoic Ice Age in Time and Space*, edited by C. R. Fielding, T. D. Frank, and J. L. Isbell, 343–354. Geological Society of America Special Paper 441.

Geslin, J. K. 1998. "Distal Ancestral Rocky Mountain Tectonism: Evolution of the

Pennsylvanian-Permian Oquirrh-Wood River Basin, Southern Idaho." *Geological Society of America Bulletin* 110:644–663.

Gehrels, G. E., and W. R. Dickinson. 2000. "Detrital Zircon Geochronology of the Antler Overlap and Foreland Basin Assemblages, Nevada." In *Paleozoic and Triassic Paleogeography and Tectonics of Western Nevada and Northern California*, edited by M. J. Soreghan and G. E. Gehrels, 57–63. Geological Society of America Special Paper 347.

Gilmore, E. H. 2007. "New Carboniferous Bryozoa of the Bird Spring Formation, Southern Nevada." *Journal of Paleontology* 81, no. 3: 581–587.

Goddéris, Y., Y. Donnadieu, S. Carretier, M. Aretz, G. Dera, M. Macouin, and V. Regard. 2017. "Onset and Ending of the Late Paleozoic Ice Ages Triggered by Tectonically Paced Rock Weathering." *Nature Geoscience* 10: 383–386.

Hiatt, E. E., and D. A. Budd. 2003. "Extreme Paleoceanographic Conditions in a Paleozoic Oceanic Upwelling System: Organic Productivity and Widespread Phosphogenesis in the Permian Phosphoria Sea." In *Extreme Depositional Environments: Mega End Members in Geologic Time*, edited by M. A. Chan and A. W. Archer, 1–20. Geological Society of America Special Paper 370.

Hofmann, R., M. Hautmann, and H. F. R. Bucher. 2016. "Diversity Partitioning in Permian and Early Triassic Benthic Ecosystems of the Western USA: A Comparison." *Historical Biology* 29, no. 7:1–13.

Joachimski, M. M., A. S. Alekseev, A. Grigoryan, and Y. A. Gatovsky. 2020. "Siberian Trap Volcanism, Global Warming and the Permian-Triassic Mass Extinction: New Insights from Armenian Permian-Triassic Sections." *Geological Society of America Bulletin* 132, nos. 1–2:427–443.

Ketner, K. B. 2008. "The Inskip Formation, the Harmony Formation, and the Havallah Sequence of Northwestern Nevada—An Interrelated Paleozoic Assemblage in the Home of the Sonoma Orogeny." U.S. Geological Survey Professional Paper 1757.

———. 2009. "Mid-Permian Phosphoria Sea in Nevada and the Upwelling Model." U.S. Geological Survey Professional Paper 1764.

Lane, H. R., L. Brenckle, J. F. Baesemann, and B. Richards. 1991. "The IUGS Boundary in the Middle of the Carboniferous: Arrow Canyon, Nevada, USA." *Episodes* 22, no. 4:272–283.

Lawton, T. F., H. Cashman, J. H. Trexler Jr., and W. J. Taylor. 2017. "The Late Paleozoic Southwestern Laurentian Borderland." *Geology* 45:675– 678.

Martin, L. G., I. P. Montañez, and J. W. Bishop. 2012. "A Paleotropical Carbonate-dominated archive of Carboniferous Icehouse Dynamics, Bird Spring Formation, Southern Great Basin, USA." *Palaeogeography, Palaeoclimatology, Palaeoecology* 329–330:64–82.

Matheson, E. J., and T. D. Frank. 2020. "An Epeiric Glass Ramp: Permian Low-Latitude Neritic Siliceous Sponge Colonization and Its Novel Preservation (Phosphoria Rock Complex)." *Sedimentary Geology* 399: article ID 105568.

Miller, R. P., and P. L. Heller. 1994. "Depositional Framework and Controls on Mixed Carbonate-Siliciclastic Gravity Flows: Pennsylvanian-Permian Shelf to Basin Transect, South-Western Great Basin, USA." *Sedimentology* 41, no. 1:1–20.

Mollazal, Y. 1961. "Petrology and Petrography of Ely Limestone in Part of Eastern Great Basin." *Brigham Young University Geology Studies* 8: 3–35.

Montañez, I. P., and C. J. Poulsen. 2013. "The Late Paleozoic Ice Age: An Evolving Paradigm." *Annual Review of Earth and Planetary Sciences* 41:629–656.

Moore, T. E., B. L. Murchey, and A. G. Harris. 2000. "Significance of Geologic and Biostratigraphic Relations between the Overlap Assemblage and Havallah Sequence, Southern Shoshone Range, Nevada." In *Geology and Ore Deposits 2000: The Great Basin and Beyond, Symposium Proceedings, May 15–18*, edited by J. K. Cluer, J. G. Price, E. M. Struhsacker, R. D. Hardyman, and C. L. Morris, 397–418. Reno/Sparks, NV: Geological Society of Nevada.

Murchey, B. L. 1990. "Age and Depositional Setting of Siliceous Sediments in the Upper Paleozoic Havallah Sequence near Battle Mountain, Nevada: Implications for the Paleogeography and Structural Evolution of the Western Margin of North America." In *Paleozoic and Early Mesozoic Paleogeographic Relations: Sierra Nevada, Klamath Mountains, and Related Terranes*, edited by D. S. Harwood and M. M. Miller, 137–155. Geological Society of America Special Paper 255.

Nelson, W. J., and S. G. Lucas. 2011. "Carboniferous Geologic History of the Rocky Mountain Region." In *Fossil Record 3*, edited by R. M. Sullivan, S. G. Lucas, and J. A. Spielmann, 115–142. *New Mexico Museum of Natural History and Science Bulletin 53*.

Payne, J. L., and M. E. Clapham. 2012. "End-Permian Mass Extinction in the Oceans: An Ancient Analog for the Twenty-First Century?" *Annual Reviews of Earth and Planetary Science* 40:89–111.

Penn, J. L., C. Deutsche, J. L. Payne, and E. A. Sperling. 2018. "Temperature-Dependent Hypoxia Explains Biogeography and Severity of End-Permian Marine Mass Extinction." *Science* 362:1–6.

Piper, D. Z., and P. K. Link. 2002. "An Upwelling Model for the Phosphoria Sea: A Permian Ocean-Margin Sea in Northwest United States." *American Association of Petroleum Geologists Bulletin* 86, no. 7:1217–1235.

Plas, L. P., Jr. 1972. "Upper Wolfcampian (?) Mollusca from the Arrow Canyon Range, Clark County, Nevada." *Journal of Paleontology* 46, no. 2:249–260.

Pommer. M., and J. Sarg. 2019. "Biochemical and Stratigraphic Controls on Pore-System Evolution, Phosphoria Rock Complex (Permian), Rocky Mountain Region, USA." *Society of Economic Paleontologists and Mineralogists (SEPM) Special Publication* 112.

Read, M. T., and M. K. Nestell. 2019. "Lithostratigraphy and Fusulinid Biostratigraphy of the Upper Pennsylvanian-Lower Permian Riepe Spring Limestone at Spruce Mountain Ridge, Elko County, Nevada, U.S.A." *Stratigraphy* 16, no. 4, p.

Rice, J. A., and D. B. Loope. 1991. "Wind-Reworked Carbonates, Permo-Pennsylvanian of Arizona and Nevada." *Geological Society of America Bulletin* 103, no. 2:254–267.

Ritter, S. M., and T. S. Robinson. 2009. "Sequence Stratigraphy and Biostratigraphy of Carboniferous-Permian Boundary Strata in Western Utah." In *Geology and Geologic Resources and Issues of Western Utah*, edited by B. T. Tripp et al., 27–42. Utah Geological Association Publication 38.

St. Aubin-Hietpas, L. A. 1983. Carbonate Petrology and Paleoecology of Permo-Carboniferous Rocks, Southwestern Millard County, Utah. *Brigham Young University Geology Studies* 30:113–143.

Saller, A. H., and W. R. Dickinson. 1982. "Alluvial to Marine Facies Transition in the Antler Overlap Sequences, Pennsylvanian and Permian of North-Central Nevada." *Journal of Sedimentary Petrology* 52:925–940.

Soregham, G. S., D. E. Sweet, K. R. Marra, C. F. Eble, M. J. Soreghan, R. D. Elmore, S. A. Kaplan, and M. D. Blum. 2007. "An Exhumed Late Paleozoic Canyon in the Rocky Mountains." *Journal of Geology* 115: 471–481.

Stevens, C. H., P. Stone, and J. S. Miller. 2005. "A New Reconstruction of the Paleozoic Continental Margin of Southwestern North America: Implications for the Nature and Timing of Continental Truncation and the Possible Role of the Mojave-Sonora Megashear." In *The Mojave-Sonora Megashear Hypothesis: Development, Assessment, and Alternatives*, edited by T. H. Anderson, J. A. Nourse, J. W. McKee, and M. B. Steiner, 597–618. Geological Society of America Special Paper 393.

Stevens, C. H. 2008. "Fasciculate Rugose Corals from the Gzhelian and Lower Permian Strata, Pequop Mountains, Northeast Nevada." *Journal of Paleontology* 82, no. 6:1190–1200.

Stevens, C. H., and P. Stone. 2007. "The Pennsylvanian-Early Permian Bird Spring Carbonate Shelf, Southeastern California: Fusulinid Biostratigraphy, Paleogeographic Evolution, and Tectonic Implications." Geological Society of America Special Paper 429.

Stone, Paul, C. H. Stevens, K. A. Howard, and T. D. Hoisch. 2013. "Stratigraphy and Paleogeographic Significance of the Pennsylvanian-Permian Bird Spring Formation in the Ship Mountains, Southeastern California." U.S. Geological Survey Scientific Investigations Report 2013-5109.

Sturmer, D., J. Trexler, and P. H. Cashman. 2018. "Tectonic Analysis of the Pennsylvanian Ely-Bird Spring Basin: Late Paleozoic Tectonism on the Southwestern Laurentia Margin and Distal Limit of the Ancestral Rocky Mountains." *Tectonics* 37, no. 2:604–620.

Sumida, S. S., G. M. Albright, and E. Rega. 1999. "Late Paleozoic Fishes of Utah." In *Vertebrate Paleontology in Utah*, edited by D. Gillette, 11–29. Utah Geological Survey Miscellaneous Publication 99-1.

Tapanila, L., J. Pruitt, A. Pradel, C. D. Wilga, J. B. Ramsay, R. Schlader and D. A. Didier. 2013. "Jaws for a Spiral-Tooth Whorl: CT Images Reveal Novel Adaptation and Phylogeny in Fossil *Helicoprion*." *Biology Letters* 9, no. 2. https://doi.org/10.1098/rsbl.2013.0057.

Theodore. T. G., I. Berger, D. A. Singer, A. G. Harris, and C. H. Stevens. 2004. "Synthrusting Deposition of the Pennsylvanian and Permian Strathearn Formation, Northern Carlin Trend, Nevada." *Sedimentary Geology* 165, nos. 1–2:1–28.

Trexler, J. H., Jr., P. H. Cashman, W. S. Snyder, and I. Davydov. 2004. "Late Paleozoic Tectonism in Nevada: Timing, Kinematics, and Tectonic Significance." *Geological Society of America Bulletin* 116, nos. 5–6:525–538.

Utah Geological Survey, and A. J. Wells. 2017. "Fossil Fusulinid Evaluation Results for the Rush Valley, Wildcat Mountain, Grouse Creek, and Tremonton 30' × 60' Quadrangles, Utah." Utah Geological Survey Open-File Report 664.

Wardlaw, B. R., and J. W., Collinson. 1986. "Paleontology and Deposition of the Phosphoria Formation." *University of Wyoming Contributions to Geology* 24, no. 2:107–142.

Wardlaw, B. R., D. M. Gallegos, V. Chernykh, and W. S. Snyder. 2015. "Early Permian Conodont Fauna and Stratigraphy of the Garden Valley Formation, Eureka County, Nevada." *Micropaleontology* 61, nos. 4–5:369–387.

Wignall, B., and R. J. Twitchett. 2002. "Extent, Duration, and Nature of the Permian-Triassic Superanoxic Event." In *Catastrophic Events and Mass Extinctions: Impacts and Beyond*, edited by C. Koeberl and K. G. MacLeod, 395–413. Geological Society of America Special Paper 356.

Wilson, E. C. 1991. "Permian Corals from the Spring Mountains, Nevada." *Journal of Paleontology* 65, no. 5:727–741.

Yancey, T. E., and C. H. Stevens. 1981. "Early Permian Fossil Communities in Northeastern Nevada and Northwestern Utah." In *Communities of the Past*, edited by J. Gray, A. J. Boucot, and W. B. N. Berry, 243–268. Stroudsburg, PA: Hutchinson Ross.

Yose, L. A., and P. L. Heller. 1989. "Sea-level Control of Mixed-Carbonate-Siliciclastic, Gravity-Flow Deposition: Lower Part of the Keeler Canyon Formation (Pennsylvanian), Southeastern California." *Geological Society of America Bulletin* 101, no. 3:427–439.

Chapter 8. The Last Gasp of the Great Basin Seafloor: The Oceans of the West in the Age of Reptiles

Balini, M., J. F. Jenks, R. Martin, C. A. McRoberts, M. J. Orchard, and N. J. Silberling. 2015. "The Carnian/Norian Boundary Succession at Berlin-Ichthyosaur State Park (Upper Triassic, Central Nevada, USA)." *Paläontologische Zeitschrift* 89:399–433.

Boyer, D. L., D. J. Bottjer, and M. L. Droser. 2004. "Ecological Signature of Lower Triassic Shell Beds of the Western United States." *Palaios* 19, no. 4:372–380.

Brayard, A., L. J. Krumenacker, J. P. Botting, J. F. Jenks, K. G. Bylund, E. Fara, E. Vennin, et al. 2017. "Unexpected Early Triassic Marine Ecosystem and the Rise of the Modern Evolutionary Fauna." *Science Advances* 3, no. 2: e1602159.

Bucher, H. 1992. "Ammonoids of the Hyatti Zone and the Anisian Transgression in the Triassic Star Peak Group, Northwestern Nevada, USA." *Palaeontographica Abt. A* 223:137–166.

Burke, D. B., and N. J. Silberling. 1973. "The Auld Lang Syne Group, of Late Triassic and Jurassic (?) Age, North-Central Nevada." *U.S. Geological Survey Bulletin* 1394-E.

Camp, C. L. 1980. "Large Ichthyosaurs from the Upper Triassic of Nevada." *Palaeontographica Abt. A* 170:139–200.

Caravaca, G., A. Brayard, E. Vennin, M. Guiraud, L. Le Pourhiet, A.-E. Grosjean, C. Thomazo, et al. 2017. "Controlling Factors for the Differential Subsidence in the Sonoma Foreland Basin (Early Triassic, Western USA)." *Geological Magazine* 155, no. 6:1–25.

Collinson, J. W., C. G. St. C. Kendall, and J. B. Marcantel. 1976. "Permian-Triassic Boundary in Eastern Nevada and West-Central Utah." *Geological Society of America Bulletin* 87(5):821–824.

Crafford, A. E. J. 2008. "Paleozoic Tectonic Domains of Nevada: An Interpretive Discussion to Accompany the Geologic Map of Nevada." *Geosphere* 4, no. 1:260–291.

Cuny, G., O. Rieppel, and P. M. Sander. 2001. "The Shark Fauna from the Middle Triassic (Anisian) of North-Western Nevada." *Zoological Journal of the Linnean Society* 133:285–301.

Dickinson, W. R. 2004. "Evolution of the North American Cordillera." *Annual Review of Earth and Planetary Sciences* 32:13–45.

———. 2006. "Geotectonic Evolution of the Great Basin." *Geosphere* 2(7):353–368.

Fröbisch, N. B., J. Fröbisch, P. M. Sander, L. Schmitz, and O. Rieppel. 2013. "Macropredatory Ichthyosaur from the Middle Triassic and the Origin of Modern Trophic Networks." *Proceedings of the National Academy of Sciences* 110(4):1393–1397.

Gabrielse, H., W. S. Snyder, and J. H. Stewart. 1983. "Sonoma Orogeny and Permian to Triassic Tectonism in Western North America." *Geology* 11(8):484.

Hogler, J. A. 1992. "Taphonomy and Paleoecology of Shonisaurus Popularis (Reptilia: Ichthyosauria)." *Palaios* 7:108–117.

Heck, F. R., and R. C. Speed. 1987. "Triassic Olistostrome and Shelf–Basin Transition in the Western Great Basin: Paleogeographic Implications." *Geological Society of America Bulletin* 99(4):539–551.

Hofmann, R., M. Hautmann, and H. Bucher. 2013. "A New Paleoecological Look at the Dinwoody Formation) Lower Triassic, Western USA): Instrinsic versus Extrinsic Controls on Ecosystem Recovery after the End-Permian Mass Extinction." *Journal of Paleontology* 87, no. 5:854–880.

Hofmann, R., M. Hautmann, M. Wasmer, and H. Bucher. 2012. "Palaeoecology of the Spathian Virgin Formation (Utah, USA) and Its Implication for the Early Triassic Recovery." *Acta Palaeontologica Polonica* 58, no. 1: 149–173.

Hogler, J. A. 1992. "Taphonomy and Paleoecology of *Shonisaurus popularis* (Reptilia: Ichthyosauria)." *Palaios* 7, no. 1:108–117.

Jenks, F. F., J. A. Spielmann, and S. G. Lucas. 2007. "Triassic Ammonoids: A Photographic Journey." In *Triassic of the American West*, edited by S. G. Lucas and J. A. Speilmann, 33–80. New Mexico Museum of Natural History and Science Bulletin 40.

Jones, D. L., and R. M. Burt. 1990. "Synopsis of Late Paleozoic and Mesozoic Terrane Accretion within the Cordillera of Western North America." *Philosophical Transactions of the Royal Society of London, Series A, Mathematical and Physical Sciences* 331, no. 1620: 479–486.

Kelley, N. P., R. Motani, P. Embree and M. J. Orchard. 2016. "A New Lower Triassic Ichthyopterygian Assemblage from Fossil Hill, Nevada." *PeerJ* 4:e1626. https://doi.org/10 .7717/peerj.1626.

Ketner, K. B. 2008. "The Inskip Formation, the Harmony Formation, and the Havallah Sequence of Northwestern Nevada—an Interrelated Paleozoic Assemblage in the Home of the Sonoma Orogeny." U.S. Geological Survey Professional Paper 1757.

Kosch, B. F. 1990. "A Revision of the Skeletal Reconstruction of *Shonisaurus popularis* (Reptilia: Ichthyosauria)." *Journal of Vertebrate Paleontology* 10, no. 4:512–514.

LaMaskin, T. A., R. J. Dorsey, J. D. Vervoot, M. D. Schmitz, K. P. Tumpane, and N. O. Moore. 2015. "Westward Growth of Laurentia by Pre–Late Jurassic Terrane Accretion, Eastern Oregon and Western Idaho, United States." *Journal of Geology* 123: 233–267.

Laws, R. A. 1982. "Late Triassic Depositional Environments and Molluscan Associations from West-Central Nevada." *Palaeogeography, Palaeoclimatology, Palaeoecology* 37, nos. 2–4: 131–149.

Leidy, J. 1868. "Notice of Some Reptilian Remains from Nevada." *Proceedings of the Philadelphia Academy of Sciences* 20:177–178.

Lucas, S. G., and M. J. Orchard. 2007. "Triassic Lithostratigraphy and Biostratigraphy North of Currie, Elko County, Nevada." In *Triassic of the American West*, edited by S. G. Lucas and J. A. Spielman, 119–126. New Mexico Museum of Natural History and Science Bulletin 40.

Lupe, R., and N. J. Silberling. 1985. "Genetic Relationship between Lower Mesozoic Continental Strata of the Colorado Plateau and Marine Strata of the Western Great Basin: Significance for Accretionary History of Cordilleran Lithotectonic Terranes." In *Tectonostratigraphic Terranes of the Circum-Pacific Region*, edited by D. G. Howell, 263–271. Houston, TX: Circum-Pacific Council for Energy and Mineral Resources, Earth Science Series 1.

Marenco. J., J. M. Griffin, M. L. Fraiser, and M. E. Clapham. 2012. "Paleoecology and Geochemistry of Early Triassic (Spathian) Microbial Mounds and Implications for Anoxia following the End-Permian Mass Extinction." *Geology* 40, no. 8:715–719.

Martindale, R. C., D. J. Bottjer, and F. A. Corsetti. 2012. "Platy Cora; Patch Reefs from the Eastern Panthalassa (Nevada, USA): Unique Reef Construction in the Late Triassic." *Palaeogeography, Palaeoclimatology, Palaeoecology* 313–314:41–58.

Mary, M., and A. D. Woods. 2008. "Stromatolites of the Lower Triassic Union Wash Formation, CA: Evidence for Continued Post-Extinction Environmental Stress in Western North America though the Spathian." *Palaeogeography, Palaeoclimatology, Palaeoecology* 261, nos. 1–2:78–86.

Mazin, J.-M., and H. F. R. Bucher. 1987. "*Omphalosaurus nettarhynchus*, a New Omphalosaurid Species (Reptilia, Ichthyopterygia) from the Spathian of the Humboldt Range, Nevada, USA." *Comptes Rendus de l'Académie des Sciences, Série 2, Mécanique, Physique, Chimie, Sciences de l'Univers, Sciences de la Terre* 305, no. 9:823–828.

McGowan, C., and R. Motani. 1999. "A Reinterpretation of the Upper Triassic Ichthyosaur *Shonisaurus*." *Journal of Vertebrate Paleontology* 19, no. 1:42–49.

Merriam, J. C. 1908. "Triassic Ichthyosauria with Special Reference to the American Forms." *Memoirs of the University of California* 1:1–196.

Monarrez, M., and N. Bonuso. 2014. "Pattern of Fossil Distribution within Their Environmental Context from the Middle Triassic in South Canyon, Central Nevada, USA." *Journal of Palaeogeography* 3, no. 1:74–89.

Monnet, C., and H. Bucher. 2005. "New Middle and Late Anisian (Middle Triassic) Ammonoid Faunas from Northwestern Nevada (USA): Taxonomy and Biochronology." *Fossils and Strata* 52.

Monnet, C., H. Bucher, M. Wasmer, and J. Guex. 2010. "Revisions of the Genus *Acrochordiceras* Hyatt, 1877 (Ammonoidea, Middle Triassic): Morphology, Biometry, Biostratigraphy and Intra-Specific Variability." *Palaeontology* 53, no. 5:961–996.

Motani, R. 2009. "The Evolution of Marine Reptiles." *Evolution: Education and Outreach* 2:224–235.

Pruss, S. B., F. A. Corsetti, and D. J. Bottjer. 2005. "Environmental Trends of Early Triassic Biofabrics: Implications for Understanding the Aftermath of the End-Permian Mass Extinction." In *Developments in Palaeontology and Stratigraphy*. Vol. 20, *Understanding Late Devonian and Permian-Triassic Biotic and Climatic Events: Towards an Integrated Approach*, edited by D. J. Over, J. R. Morrow, and P. B. Wignall, 313–332. Elsevier Science.

Pruss, S. B., and J. L. Payne. 2009. "Early Triassic Microbial Spheroids in the Virgin Limestone Member of the Moenkopi Formation, Nevada, USA." *Palaios* 24, no. 2:131–136.

Rieppel, O., P. M. Sander, and G. W. Storrs. 2002. "The Skull of the Pistosaur *Augustasaurus* from the Middle Triassic of Northwestern Nevada." *Journal of Vertebrate Paleontology* 22, no. 3:577–592.

Romano, C., J. F. Jenks, R. Jattiot, T. M. Scheyer, K. G. Bylund, and H. Bucher. 2017. "Marine Early Triassic Actinopterygii from Elko County (Nevada, USA): Implications for the Smithian Equatorial Vertebrate Eclipse." *Journal of Paleontology* 91:1025–1046.

Romano, C., I. Kogan, J. F. Jenks, I. Jerjen, and W. Brinkmann. 2012. "*Saurichthys* and Other Fossil Fishes from the Late Smithian (Early Triassic) of Bear Lake County (Idaho, USA), with a Discussion of Saurichthyid Palaeogeography and Evolution." *Czech Geological Survey Bulletin of Geosciences* 87, no. 3:543–570.

Romano, C., A. Lopez-Arbarello, D. Ware, J. F. Jenks, and W. Brinkmann. 2019. "Marine Early Triassic Actinopterygii from the Candelaria Hills (Esmeralda County, Nevada, USA)." *Journal of Paleontology* 93, no. 5: 971–1000.

Roniewicz, E., and G. D. Stanley. 2013. "Upper Triassic Corals from Nevada, Western North America, and the Implications for Paleoecology and Paleogeography." *Journal of Paleontology* 87, no. 5:934–964.

Saleeby, J. B., C. Busby-Spera, J. S. Oldow, G. C. Dunne, J. E. Wright, D. S. Cowan, N. W. Walker, and R. W. Allmendinger. 1992. "Early Mesozoic Tectonic Evolution of the Western U.S. Cordillera." In *The Cordilleran Orogen: Conterminous U.S.: The Geology of North America*, edited by B. C. Burchfiel, P. W. Lipman, and M. L. Zoback, 107–168. Decade of North America, vol. G-3. Boulder, CO: Geological Society of America.

Sander, P. M., and C. Faber. 2003. "The Triassic Marine Reptile Omphalosaurus: Osteology, Jaw Anatomy, and Evidence for Ichthyosaurian Affinities." *Journal of Vertebrate Paleontology* 23, no. 4:799–816.

Sander, M., O. C. Rieppel, and H. Bucher. 1994. "New Marine Vertebrate Fauna from the Middle Triassic of Nevada." *Journal of Paleontology* 68, no. 3:676–680.

———. 1996. "A New Pistosaurid (Reptilia: Sauropterygia) from the Triassic of Nevada and Its Implications for the Origin of the Plesiosaurs." *Journal of Vertebrate Paleontology* 17, no. 3:526–533.

Sandy, M. R., and G. D. Stanley. 1993. "Late Triassic Brachiopods from the Luning Formation, Nevada, and Their Palaeobio-

geographical Significance." *Palaeontology* 36, no. 2:439–480.

Scheyer, T. M., C. Romano, J. Jenks, and H. Bucher. 2014. "Early Triassic Marine Biotic Recovery: The Predators Perspective." *PLoS ONE* 9, no. 3: e88987.

Schmitz, L., and P. M. Sander. 2002. "Phylogenetic Implications of New Mixosaur (Ichthyosauria) Material from the Middle Triassic of Nevada (USA)." *Schriftenreihe der Deutschen Geologischen Gesellschaft* 21:299–300.

Schmitz, L., P. M. Sander, G. W. Storrs, and O. Rieppel. 2004. "New Mixosauridae (Ichthyosauria) from the Middle Triassic of the Augusta Mountains (Nevada, USA) and Their Implications for Mixosaur Taxonomy." *Palaentographica A* 270, nos. 4–6:133–162.

Senobari-Daryan, B., and G. D. Stanley Jr. 1992. "Late Triassic Thalamid Sponges from Nevada." *Journal of Paleontology* 66, no. 2: 183–193.

Silberling, N. J., and K. M. Nichols. 1982. "Middle Triassic Molluscan Fossils of Biostratigraphic Significance from the Humboldt Range, Northwestern Nevada." U.S. Geological Survey Professional Paper 1207.

Silberling, N. J., and R. E. Wallace. 1969. "Stratigraphy of the Star Peak Group (Triassic) and Overlying Lower Mesozoic Rocks, Humboldt Range, Nevada." U.S. Geological Survey Professional Paper 592.

Silberling, N. J., D. L. Jones, J. W. H. Monger, and P. J. Coney. 1992. "Lithotectonic Terrane Map of the North American Cordillera." U.S. Geological Survey Miscellaneous Investigations Series Map I-2176, scale 1:5,000,000.

Snyder, S. S., and H. K. Brueckner. 1983. "Tectonic Evolution of the Golconda Allochthon, Nevada: Problems and Perspectives." In *Paleozoic and Early Mesozoic Rocks in Microplates of Western North America*, edited by C. A. Stevens, 103–123. Society of Economic Paleontologists and Mineralogists, Pacific Section.

Song, H., et al. 2018. "Decoupled Taxonomic and Ecological Recoveries from the Permo-Triassic Extinction." *Science Advances* 4, no. 10:1–6.

Speed, R. C. 1979. "Collided Paleozoic Microplate in the Western United States." *Journal of Geology* 87:279–292.

Sperling E. A., and J. C. Ingle Jr. 2006. "A Permian-Triassic Boundary Section at Quinn River Crossing, Northwestern Nevada, and Implications for the Cause of the Early Triassic Chert Gap on the Western Pangean Margin." *Geological Society of America Bulletin* 118: 733–746.

Stanley, G. D., Jr. 2005. "Coral Microatolls from the Triassic of Nevada: Oldest Scleractinians Examples." *Coral Reefs* 24:247.

Stanley, G. D., Jr., and K. P. Helme. 2010. "Middle Triassic Coral Growth Bands and Their Implication for Photosymbiosis." *Palaios* 25: 754–762.

Stewart, J. H. 1997. "Triassic and Jurassic Stratigraphy and Paleogeography of West-Central Nevada and Eastern California." U.S. Geological Survey Open-File Report 97-495.

Stewart, J. H., T. H. Anderson, G. B. Haxel, L. T. Silver, and J. E. Wright. 1986. "Late Triassic Paleo-Geography of the Southern Cordillera: The Problem of a Source for Voluminous Volcanic Detritus in the Chinle Formation of the Colorado Plateau Region." *Geology* 14:567–570.

Stewart, J. H., F. G. Poole, and R. F. Wilson. 1972. "Stratigraphy and Origin of the Triassic Moenkopi Formation and Related Strata in the Colorado Plateau Region." U.S. Geological Survey Professional Paper 691.

Stubbs, T. L., and M. J. Benton. 2016. "Ecomorphological Diversifications of Mesozoic Marine Reptiles: The Roles of Ecological Opportunity and Extinction." *Paleobiology* 42, no. 4:547–573.

Sues, H.-D., and J. Clark. 2005. "Thallatosaurs (Reptilia: Diapsida) from the Late Triassic (Carnian) of Nevada and Their Paleobiogeographic Significance (abst)." *Journal of Vertebrate Paleontology* 25, supplement to no. 3:119A.

Tackett, L. S., and D. J. Bottjer. 2016. "Paleoecological Succession of Norian (Late Triassic) Benthic Fauna in Eastern Panthalassa (Luning and Gabbs Formations, West-Central Nevada)." *Palaios* 31, no. 4:190–202.

Van der Meer, D. G., T. H. Torsvik, W. Spakman, D. J. J. van Hinsbergen, and M. L. Amaru. 2012. "Intra-Panthalassa Ocean Subduction Zones Revealed by Fossil Arcs and Mantle Structure." *Nature Geoscience* 5:215–219.

Vennin, E., N. Olivier, A. Brayard, I. Bour, N. Thomazo, G. Escarguel, E. Fara, K. G. Bylund, J. F. Jenks, D. A. Stephen, and R. Hofmann. 2015. "Microbial Deposits in the Aftermath of the End-Permian Mass Extinction: A Diverging Case from the Mineral Mountains (Utah: USA)." *Sedimentology* 62: 753–792.

Wallace, R. E., D. B. Tatlock, and N. J. Silberling. 1960. "Intrusive Rocks of Permian and Triassic Age in the Humboldt Range, Nevada." U.S. Geological Survey Professional Paper 400B:B291–B293.

Waller, T. R., and G. D. Stanley. 2005. "Middle Triassic Pteriomorphian Bivalvia (Mollusca) from the New Pass Range, West-Central Nevada: Systematics, Biostratigraphy, Paleoecology, and Paleobiology." *Journal of Paleontology* 79, issue S61:1–58.

Ware, D., J. F. Jenks, M. Hautmann, and H. Bucher. 2011. "Dienerian (Early Triassic) Ammonoids from the Candelaria Hills (Nevada, USA) and Their Significance for Palaeobiogeography and Palaeoceanography." *Swiss Journal of Geosciences* 104:161–181.

Woods, A. D., and D. J. Bottjer. 2000. "Distribution of Ammonoids in the Lower Triassic Union Wash Formation (Eastern California): Evidence for Paleoceanographic Conditions during Recovery from the End-Permian Mass Extinction." *Palaios* 15, no. 6:535–545.

Woods, A. D. 2013. "Microbial Ooids and Cortoids from the Lower Triassic (Spathian) Virgin Limestone, Nevada, USA: Evidence for an Early Triassic Microbial Bloom in Shallow Depositional Environments." *Global and Planetary Change* 105:91–101.

Woods, A. D., P. D. Alms, P. M. Monarrez, and S. Mata. 2019. "The Interaction of Recovery and Environmental Conditions: An Analysis of the Outer Shelf Edge of Western North America during the Early Triassic." *Palaeogeography, Palaeoclimatology, Palaeoecology* 513:52–64.

Wyld, S. J. 1991. "Permo-Triassic Tectonism in Volcanic Arc Sequences of the Western U.S. Cordillera and Implications for the Sonoma Orogeny." *Tectonics* 10, no. 5:1007–1017.

———. 2000. "Triassic Evolution of the Arc and Backarc of Northwestern Nevada, and Evidence for Extensional Tectonism." In *Paleozoic and Triassic Paleogeography and Tectonics of Western Nevada and Northern California,* edited by M. J. Soreghan and G. E. Gehrels, 185–207. Geological Society of America Special Paper 347.

Index